Biodynamic Farming a

Ehrenfried E. Pfeiffer

BIODYNAMIC
FARMING AND GARDENING

Renewal and Preservation of Soil Fertility

EHRENFRIED E. PFEIFFER

Portalbooks ≈ 2021

Portalbooks

An imprint of SteinerBooks / Anthroposophic Press
402 Union Street, No. 58, Hudson, NY 12534
www.steinerbooks.org

Design: Jens Jensen

1st edition 1938

2nd revised edition 1940

3rd revised edition 1943

4th revised edition 2021

LIBRARY OF CONGRESS CONTROL NUMBER: 2020946669

ISBN: 978-1-938685-29-3 (paperback)
ISBN: 978-1-938685-30-9 (eBook)

Contents

List of Illustrations

Introduction

Every agricultural enterprise
is a self-contained, biological unit

To the degree that human beings participate in nature through nourishment, they will likewise, as producers and consumers, be interested in the fruitfulness of the earth. This is especially true of farmers, because for them the yield of the earth represents their livelihood, while the cultivation of the soil is their occupation. For consumers, the quality and taste of the farm or garden's yield offer the corporal basis of life and health.

The methods of agriculture are not, therefore, the center of a mere occupational life, but rather the focal point of a common human cultural life. Culture, then, in its original sense, means work performed on the earth, just as in the broadest sense it means an achievement of the human spirit. Accordingly, at a high cultural level, a people has well-cultivated farms and gardens, as well.

Modern agriculture has become an economic objective; it has been invaded by economic thinking. Increasing industrialization and technical considerations now exercise an influence on the structure of the agricultural establishment. The soil's productive capacity and the cost involved are the basis for evaluation of the profit motive. Likewise, scientific research has investigated the nutritive content of the soil, as well as the nutritive balance between soil and plant. Comprehensive studies have been made of the soil's mineral content. Under the influence of a progressive chemical science since the middle of the last century, so-called rational farmers have been forced increasingly to assume the role of small manufacturers, whose means of production are the soil, their farming implements and tools, and the growth characteristics of certain plants.

Just as the usefulness of a motor can be calculated, similarly the usefulness of agriculture is calculable. Agriculture, however, is far less amenable to quantification. A great deal of effort must be expended to obtain even a modest return. When economic depressions occur in world commerce, this return is even smaller—that is, the agricultural "production machine" pays poorly. If the production of an industrial machine does not pay, the manufacturer shuts down; the means of production are sold, converted to other use, or liquidated entirely.

Because of such limitations of modern industrial economy, its mechanistic ways of thinking cannot be applied to agriculture. A single abandoned farm in a well-cultivated region means widespread damage to the environment—for example, the spread of weed seeds by the wind and a change in soil quality. The presence of many such farms together in the same area can lead to the devastation of a whole district and to natural catastrophes. Examples of this sort are known; in ancient times, Mesopotamia and, more recently, dust storms in the U.S. In matters of technique, we are dealing with a preponderance of inorganic material, for which the calculability and individual factors are clearly definable. Agriculture, however, deals with the conditions of life. The growth, health, and disease of plants and animals, as well as restoration of the soil, are continuously variable factors whose individual levels of importance are interdependent and unite into *a higher unity, or a complete whole as an organism.*

The technological process of production takes raw materials and manufactures them into finished products. The production machine between these two conditions alters very little, except through the wear and tear to which it is subjected. Agriculture receives fertilizer and seed as "raw materials" and delivers vegetables, grains, beets, and other produce as the "finished articles." The life process occurs between those two factors. Economic thinking in agriculture cannot be justified unless it includes the life process in its calculations. If biological unity of the agricultural enterprise becomes the basis of calculation, then the following statement will apply: *Whatever is biologically correct is also the most economically profitable.* Three fundamental characteristics become clear in every life phenomenon:

The *first* is the fact that life (if the corresponding conditions are at all prevalent) always wants to create growth and increase. Unlimited exuberance of growth is an unrestrained expression of life.

The *second* characteristic points to an inner condition of strain—for wherever life and building-up exists, deterioration, breakdown, decay, and death are also present. One contains the other within itself—indeed, presupposes it. Goethe beautifully expressed this reciprocal relationship in his essay on Nature: *"Life is Nature's most beautiful discovery, and Death is her artifice for obtaining abundant life."* This condition of strain between two processes is often called "equilibrium." This does not mean a stable, fixed mechanical equilibrium, but an extremely active, mobile state. This state can be better described in the sense of Hippocrates and Heraclitus as *eucrasia*—the right mixture. Thus, it means active occurrences, whose final result is "life." All of the factors that participate in the creation of a phenomenon of life stand in a certain equalizing relationship to one another. Thus, it is not merely a matter of the inner characteristics of an individual living being, but one of the influence of the whole surrounding world. This can never be over-estimated. When the harmony is disturbed, this expresses itself in a continuous dislocation of all the conditions of life.

The *third* fundamental characteristic can be outlined somewhat with the following: *The whole is not the mere sum of all its parts, but a harmonic unity of a higher order that, as organic being—as an organism with laws of a higher order—lifts the world of the physiochemical inorganic to the world of the organic and living.* Expressed consciously for the first time by Goethe, today this truth increasingly affects all of our biological thinking. Through this, we learn not only that any person, animal, or plant is an organism, but also that the cohabitation of plant world and earth, of plant with plant in certain groupings, of plant with animal and human, likewise forms itself into an organic unity. Indeed, the entire development of the "living space," let's say, of a people or a continent, is fashioned according to the same fundamental laws of "the will to evolve," or "the proper interaction of all factors," and "the organic unity of a higher order." *The disturbance of one factor means the disturbance of the whole system.* Since we are then dealing with a changeable, unstable, strained condition, an insignificant disturbance might, in due time, lead to serious consequences.

These three fundamental characteristics of life point to one result—the inner efficiency of the organism. In technological terms, we speak of a "safety factor" or a "modulus of elasticity." A steel rod breaks under a certain stress or loses its elasticity under a certain tension. Every material has a specific factor of safety that cannot be exceeded without causing structural damage. Today, we hear much about the use of technology in life— for example, the increased use of technology in agriculture. This is conceived as increased efficiency through the use of machines as means of increased production and such. All of this, however, advances only to a certain limit—namely, the biological efficiency of a given natural foundation.

The purpose of the following chapters is to show how biological unity of these agricultural efforts can be attained. Here, the author relies on many years of experience. The practical insight that comes from the management of one's own farm, as well as from the contact with the management of farms in nearly all the countries of Europe, represents the basis of one's experience. This has been broadened by journeys to North America, North Africa, and Anatolia for the purpose of observation and study.

This is how I reached a comprehensive understanding of the possibilities of the so-called biodynamic agricultural method. The originator of this method, Rudolf Steiner, provides the basis for this book. To him, first, our thanks are due, as well as to all those who, in the past 12 years, took up the impulse given by Steiner and who, in fact, have realized it in practice, whether as farmers, gardeners, foresters, or scientific researchers; their number already exceeds a thousand. Their experiences were collected by means of testimonials and the exchange of opinions.

In various sections of this book, measures for practical means of carrying out these methods—including the treatment of manure, healthy farm management, and so on—are treated in minute detail.

In conclusion, the author would like to express a more general human view. He is convinced that the cleverest methods of technology and chemistry alone do not suffice to make good farmers, even when they have mastered them well. It is a peculiarity of the farming vocation to deal with "living matter." Our entire inner attitude has to take this fact into consideration. Technology and our attitudes must be brought into harmony before improvement seems possible. I am aware of the fact that this statement will often be set aside as impractical or impertinent to the problems of Western countries; idealism doesn't not pay the bills. Although life, health, and lack of health can be expressed numerically, they are far from being something one can buy and sell. We must bear in mind the fact that we have here a creative task to perform. First, however, to create something new requires a building plan; we must have the idea for this if we don't want to be taught by the damage that arises from uncontrolled empiricism.

The reader will find references to certain "preparations"—numbers 500 to 508—that are an essential of biodynamic farming and have become known to farmers by these numbers. Certain of these preparations are used in the manure and compost piles to hasten the rotting process and to give it the proper direction. Two of the preparations are sprays for soil and plants. Biodynamic farmers can make these preparations themselves or obtain them from other biodynamic farmers. The substances from which they are made are described in this book.

Although those who wish to start biodynamic farming might have these preparations and instructions for their use, directions for *making* them are reserved to bona fide biodynamic farmers of standing. This seeming secrecy is observed to prevent these preparations from being commercialized or being manufactured by incompetent individuals. No biodynamic farmer (or anyone else) is permitted to profit financially from them.

There are now biodynamic farms in nearly all English-speaking countries. In some places, associations of biodynamic farmers have been formed to discuss problems and mutual assistance to make these preparations available.

The experiences of biodynamic farmers in all fields have been gathered at the end of this book. Thus, readers will not be burdened with unnecessary quotations and references in the text.

Those who are seriously interested in the perspective presented here will be able to obtain further information—especially advice and counsel on the method of farm conversion—from the authorized information bureaus for biodynamic methods of agriculture. For Europe, the addresses of these can be obtained from the Central Bureau of Biodynamic Agricultural Methods, Dornach near Basel, Switzerland. For the United States and Canada, from the Biodynamic Farmers and Gardeners Association, Kimberton Farms, Phoenixville, Pennsylvania.*

Ehrenfried Pfeiffer
Dornach, February 1938

---

*   Currently: Biodynamic Association; www.biodynamics.com.

CHAPTER I

The Farmer, Yesterday and Today

True farmers are on the point of disappearing. Looking at the past few centuries, we can follow the decline, step by step. Traditional customs are no longer understood and practiced, and the problems have changed. This can be described in a few sentences. In times gone by, after a week of hard work, farmers would walk through their fields on Sunday, with a heavy, swinging gait acquired by walking behind a plow. It indicates one's train of thought, expressing a deep penetration into the processes of nature. On such Sunday walks, the creative work of the week was thought through in terms that can almost be compared to the biblical epic of creation.

Farmers were often accompanied by a son who had been initiated into the mysteries of nature. Farmers described in simple words to their offspring the ways of tilling the earth and the art of sowing. Such experiences had been handed down by their ancestors from time immemorial. Fixed rules did not yet exist; they were guided only by experience. Through observation and tradition, farmers were able to use the course of nature with its signs and symbols like an almanac. The budding of this or that bush indicated the time to prepare the seed furrow. Wild plants were guides for the right moment to do this thing or that. This instinctive certainty of traditional farmers prompted them to take the necessary measures at the right moment by observing nature's course; this instinct has been lost. Uncertainty has arisen, and now a successful neighbor is often the only guide for a farmer's work. "I see a farmer on the hill today, out with the plows," says the farmer. In this way, we look to others for farming wisdom.

In previous times, it was customary to rotate crops, with intervals during which the land was allowed to lie fallow to rejuvenate the soil. During their Sunday walk, farmers might say: *When this field lies fallow for the third time, we will have the dowry ready for our daughter. Once this cow calves for the last time, our son will be ready to leave school.* Diligence and orderliness assured a safe future. A certain feeling of comfort and satisfaction was spread over the entire field of farming activities. Such an atmosphere has often been described in works of poetry.

Farmers of that older time also had a refined feeling for meteorological conditions. They felt in their bones (even without the aches of rheumatism) every change of weather and were able to adjust their work accordingly. All these semiconscious treasures of knowledge of past times can be found today only among "old eccentrics."

Scientific agriculture has decidedly altered the ways of ancient farmers. It has told farmers to abandon their old superstitions; that they can grow better crops by turning their fields into crop factories. The economic development of the twentieth century transformed farmers into agriculturists who have to "calculate" costs and output. Concepts such as "profitableness," born from the decline of ancient agricultural traditions, have become their "daily bread." Thoughts such as the following now fill the farmer's mind: *How much will this work cost? How much will this field produce? Will it pay to till this field again?* Many similar problems occupy his thoughts.

Although farmers were told that, using scientific agricultural methods, they could double the yield of their farms, the fact remains that, after years of scientific help, present-century farmers are discovering that their "doubled" yields today are no better than the single yields of previous times.*

The farmer's work is not dependable. In previous times, farmers knew that if they sowed the seed at a given time according to the moon phases, they would have a dependable crop. Now such a process is considered nonsense. Unfortunately, modern science has not developed exact rules that can replace the old farming wisdom that to indicate the time and manner of sowing. Farmers constantly have to complain about the uncertainty of their crops; they can no longer predict with certainty the course of their growth. Earth and plants have become more sensitive and erratic. Traditional farmers tell how things have changed, and, because of economic pressures today, farmers don't have time to experiment.

The German national economist Werner Sombarth says:

> The calculation of "profitableness" is an invention of the devil, whereby he fools human beings. Much of our misery is connected with its spread. It has destroyed a colorful world and transformed it into the grey and gloomy monotony of money valuation.... The *principle* of profitableness can be distinguished from the *calculation* of profitableness.... Only here, the diabolical thought hidden behind the calculation of profitableness is completely unveiled.

Thus, the "farmer" has become an "agriculturist." Development, under the pressure of "profitableness," has forced farmers to resort to using machinery to replace more expensive human labor. For a certain time, relief was seemingly obtained by this method, especially through the use of mechanical harvesters. However, in the case of machinery devised to till soil, this "relief" has been only partial. Here, the value of intensive manual labor has proved itself

---

* Compare: *Die chemische Industrie,* (no. 2, 1933. p. 166. Average use of nitrogen as fertilizer in 1,000 tons, 1912–14:-170; 1927–29:408; 1933–35:387. Average yield of 100 pounds per acre (for summer barley): 1912–14:-21.3; 1927–29:-20.0; 1933–35:-19.8. The figures given here are somewhat similar for oats.

indispensable; it cannot be replaced by machines. It is true that machines accomplish their results more quickly, but also in a more superficial way. The fine structure of the soil that once existed cannot be produced or preserved by machines. And, if it is attempted by using a "soil pulverizer," for example, then the earthworms—creators of natural humus—are destroyed. The machine age has brought about great changes in the work of those whose business is to till the soil. It has created farmers who are "growth industrialists" and "growth mechanics."

This age has also seen the ever-increasing use of mineral fertilizers. At the same time, the yield of crops that rose at first has ceased to climb at the same rate with continued use of such fertilizing methods. The situation today shows that, compared to the time prior to the [First World] War, three times as much nitrogen is being used, whereas the average yield per acre has not increased. And in some areas it has actually decreased. Nonetheless, the idea that there might be an uneconomical principle behind the fertilizing methods currently in use is often considered heretical. Yet another problem of our time is an increase in the phenomena of degeneration—i.e., plant diseases and insect pests. A description of these phenomena is unnecessary; they are a part of the daily experience of farmers.

We need only ask ourselves: *How many of our working hours today must we devote to spraying against pests and seed baths? And how many hundreds of pounds of the yield are destroyed by pests after the harvest is gathered?* Think of the role of the maize weevil. Experts tell us that we lose half of all that starts to grow. The final result of all this is that crop costs are higher today, but without a higher crop yield than 30 or 40 years ago.

Recognizing these facts leads to many questions in the minds of farmers. If they were fully alive to the seriousness of their situation, they might say to themselves as they walk their fields: *What has happened? How can I stop this unhealthy development? I am seeing the whole earth becoming sick. This earth sickness cannot be halted by the isolated measures of a single farm.*

The ways and means for regenerating a farm can be found only in a comprehensive view of the earth as an "organism," a living entity. It is clear that the methods of so-called scientific agriculture, with its goal of ever-increasing yields and its continued use of ever-more mechanical aids, has led to an impasse.

CHAPTER 2

The Situation of Agriculture

We don't need to look back to classical antiquity to study the transformation of the most fruitful land into deserts because of false or one-sided utilization. The erosion and dust storms of the American Midwest serve as a dramatic example for today. Adjoining the fruitful black soils of the Missouri–Mississippi Basin, a broad plain extends perhaps five or six hundred miles westward. cultivated land borders that plain, followed by centuries-old prairie land. To the west, this region is bounded by the Rocky Mountains. Once the prairie was home to wild bison, and then a cattle-ranching region covered by the heavy sods of prairie grass.

Because of unbalanced overgrazing and greedy tillage of the soil, along with a lack care to protect it, the sod was gradually loosened and the protective soil cover destroyed. Drought and wind then accomplished their task, and the topsoil began to "wander." In the cultivated sections, intensive soil-depleting grain culture—without beneficial crop rotation and harrowing, without rolling at the right time, and many other errors in tillage—have contributed to the loss of the soil's organic substance, at one time so rich in humus. The topsoil disintegrated and the capillary ability of the soil disappeared, further advancing erosion.

Experiments at Missouri State University, Columbia, have shown that the nitrogen and humus content of the soil, in contrast to the original prairie, has fallen about 35%. In approximately thirty years, the fertile condition of the soil has changed to such a degree that, according to Professor Jenny, only soil acidity has increased during this period.* Today it would cost more than the land is worth to return it to its original state artificially—that is, in practical terms, such restoration is no longer feasible. The dust storms that have been prevalent in recent years are completing the process of this retrogression of fertility. Many other states are seriously affected and thus a third of the cultivated area of the United States of America is on the way to becoming useless. In addition to this, the region devoured by dust storms moves eastward at the rate of about 40 miles a year.

In addition to these natural catastrophes, economic difficulties have overtaken agriculture. American farmers are faced with the dilemma either of employing as few workers as possible or abandoning their farms altogether. Human labor, because it is too expensive, has been largely replaced by machines. But mechanically working the soil can never conserve humus as effectively as manual human tillage. Soil is now hand-tilled only insofar as absolutely necessary. The biologically beneficial form of a balanced, diversified farm—with heavy legume plantings, meadows, and green manuring—was given up in favor of one-sided cultivation. The result of this course—policies lacking in foresight, as well as a few years of poor weather conditions—is that one of the great grain-producing areas of the world, the United States, has imported cattle feed for the last two years [prior to 1933], and that in 1936 this wheat-exporting country actually imported wheat.

Conditions are similar in Canada, where a 50% harvest was recently reported. If such conditions continue, it is within the bounds of possibility that two of the most important grain producing countries of the world will soon be living "hand to mouth."

If we look at European conditions, we see that there, too, we can hardly speak of bumper crops of wheat. Sir Merrick Burell,

---

* "Soil Fertility Losses under Missouri Conditions," Columbia, MO, 1933 Bul. 324, Agr. Experiment Station.

chairman of the Standing Committee of the Council of Agriculture for England, presented the following resolution in July, 1936:

> The Council of Agriculture for England is seriously concerned over the position of the national food supply, which is obviously at present one of the weakest links in the chain of National Defense. Both the fertility of the soil and the means for increased production of foodstuffs are today less than they were in 1914 and subsequent years, when the shortage of food supplies placed the country in a most perilous position. The Council, therefore, desires to urge on the government and, through it, the country regarding the great national importance of this question. It suggests that those responsible for defense should be requested to give it their immediate attention so that remedial measures may be at once set on foot.

The resolution also warned the Cabinet that the lowered reserves of fertility in the greater part of the soils of the country and the fewer employees today made a rapid expansion of the production of food absolutely impossible. Only a carefully thought-out, long-term agricultural policy, embracing all sections of the industry, would be likely to prove adequate.* The war departments in various countries often have a broader view of the agricultural realities than do the agricultural experts.

In the mid-1800s, agriculture flourished in England, whereas today there are broad areas covered with heather and ferns as soil acidity continues to increase. When we consider this carefully, we find that in principle here, too, the conditions are no different from those in the West. We can observe how, under the influence of a moist climate, the pastures and meadows are taking on an increasingly mossy character. Social and economic conditions are also doing their part to promote one-sided cultivation of the soil. It is beginning to dawn on those concerned that one-sided agriculture can produce only temporary results.

---

* *Journal of the Ministry of Agriculture*, vol. 43. no. 4, July, 19, 1936, p. 372, "Report from the Council of Agriculture for England."

There was one picture in England that especially impressed this author. He visited a farm with heavy clay soil. Analysis of the soil was discussed, and a complete lack of calcium was noted. According to the rules of the mineralizing theory, these fields should have been immediately spread with lime, but the peculiarity of the situation was that the surface of the soil (and such cases are not rare) was from 1½ to 2 feet deep, under which was a deposit of nearly pure limestone. The top had separated completely from the subsoil, continuing its separate existence like a thick rind. There were no longer any biological connections among the life of the plant roots, the soil bacteria, and the earthworms. One-sided measures of cultivation had brought about this isolation. Earthworms usually go down as far as 6 feet and more into the ground, and carry lime up from the lower levels provided they receive the proper stimulus for the formation of humus. Moreover, the roots of legumes, when these were planted, penetrated deeply into the earth and, through their capacities for making the soil elements available—capacities that are 40 to 60 times greater than those of the grains—would bring about and stimulate an interchange of elements. A properly timed aeration would also help such a process of healing, provided the first steps toward the formation of humus substances were taken.

Similarly, processes causing a continuous reduction of soil fertility are taking place all over the European continent. If we turn to the Far East, we see that, although mechanized farms represent the typical picture of the West, the opposite is the case in the East—in Northeast China and the great river valleys. Overpopulated farm areas are common there. The United States has 41 inhabitants per square mile, but Switzerland has about 225 to the same area, Germany 343, and England 660. Censuses of the heavily populated regions of China and Japan show 1,750 to 2,000 inhabitants per square mile. Of course, this is the case only in purely agricultural regions.

In the province of Shantung, a family of 12 plus a donkey, cow, and two pigs is normal for a farm of 2½ acres. In Central Europe a peasant farm of 40 acres, about 16 times the size of the Chinese farm just given, can barely support a family. In the thickly settled Asiatic regions, 240 people—in addition to their domestic animals—live on the same amount of ground and what it produces. Clearly, the extraordinarily low requirements of Eastern farmers and agricultural laborers make this possible. Nonetheless, from a biological perspective, the productivity of soil that has been able to give complete nourishment to its inhabitants for thousands of years is an amazing phenomenon. In Europe, we face the fact that soil under cultivation little more than a few centuries is no longer capable of nourishing its population completely in every respect.

Intensive Chinese soil cultivation rests on a humus and compost economy, carried on with nearly religious zeal. Everything that can be turned into soil is composted—plants, all sorts of refuse, the muck of streams, and ordinary dirt are all arranged in layers, watered, and quickly turned into humus earth. All work in China is manual labor. This conserves the soil and permits inner aeration and mixing. Mixed cultures of as many as six different kinds of plants in various states of growth utilize the interaction of plant groups with one another. To enlarge the surface area, everything is planted between and on ridges. Mineral fertilizing is still unknown there—fortunately for the Chinese. Scientists who have visited and studied such regions from their background of technical knowledge say that a crop failure in this soil is a rare phenomenon.

Here, then, it has been possible to keep a land in its original state of fertility by using the oldest cultural methods of humanity—humus conservation and manual labor. Yet, the unnatural overpopulation shows that there, too, the biological balance is disturbed. A visitor to the great river valley plains can observe on adjacent hills and mountain chains poorly grassed or steppe-like impoverished land, rapidly becoming deserts. The overpopulated fruitful regions are directly adjacent to the extreme opposite: infertile desert lands.

We know that it was a lack of food, caused by the overcrowded conditions of their country, that really drove the Chinese to cherish ancient, tested methods of cultivation with religious reverence. The value of the ground to the individual who inherits it can be seen in many laws and regulations. The land is the property of the state—formerly of the emperor. Those who cultivate the land well can claim it as their own as long as they cultivate it. When they no longer do this, another person or the state can take control of the field. If one of the many floods carries away a piece of land from the river bank, farmers have a right to "follow" it. If they are able to locate the spot where the soil and mud have been redeposited, they can settle there. If such misfortune happens in a settled region, the earlier owners and the owners who follow the earth carried down by the flood must divide their small piece of land.

When the soil of a field begins to show signs of exhaustion, the whole family cooperates by carrying topsoil in baskets and carts onto the farm, where the earth is carefully mixed with manure and plant refuse and composted, and then turned over many times. Meanwhile the lower soil of the field is reinvigorated by planting legumes. After a while, the regenerated earth is again returned to the field and the cycle of thousands of years begins again.

If we can speak in the West of "natural" catastrophes that threaten regions of human habitation, we must similarly point to the East, where other human catastrophes of the most tremendous magnitude are in the process of preparation. The overpopulated land in China has already reached the extreme limit of its productive capacity to meet even the minimal standards of existence for those living on it. Next to this land are poor, hilly regions stripped bare of trees. The process of deforestation in China begun on great stretches of land as early as 100 BC, and was nearly complete by AD 1388. The only exceptions have been certain southern and southwestern regions that, in part, have a nearly subtropical climate. The general conditions of a country's water supply are largely disturbed once the balancing effect of wooded areas is missing; without such wooded areas, the extremes of climate draw still further apart. Sudden torrents of rain and floods alternate violently with periods of drought, heat, and extreme cold. The regulating balance, water retention, cooling, and warming provided by the forests are missing.

The following is quoted from the exceptional description by Prof. G. Wegener in an article on China in the *Handbuch der Geographischen Wissenschaft,* 1936:

> On slopes of the hills, we see either a dusty, poor stand of plants of secondary quality or the complete and fearful bareness and emptiness of rocky ground. The higher portions are badly furrowed and torn by the rain, whereas the lower are always newly covered with streams of mud that destroy the fruitful soil at the base of the hills. The levels and valley floors are cultivated as gardens right up to their extreme edges. The great lack of fuel drives the inhabitants to a ruthless interference with natural reforestation, even on uncultivated lands. Richtofen recounts with a sort of rage how in North China there is widespread use of a kind of claw-like utensil, by means of which even the most modest little plant root is torn out of the soil. In some sections of the country, grass roots, dung, and so on today must already serve as the only fuel. In South China, although this was settled much later by the Chinese and is even today much more thinly populated, the exhaustion of the forests is also already far advanced and continues at a great pace, and here, too, with all its devastating results. This is especially true in the area of thickly populated sections, as well as in the region on the Xi River Delta. Chinese culture, which in its conservatism strongly remains a culture of wood, needs this wood in large amounts—and thus the consumption of forests goes ceaselessly on. It thus becomes clear that, even for the inexplicably cheerful Chinese farmers, the minimum standards of existence are being quickly left behind, and that precisely from these provinces comes the main stream of the tremendous

Chinese emigration, which Manchuria has absorbed in recent years. The denuded hills and mountainous regions, about 50% of the total area, remain practically unused.

From here, as well as from the similarly over-settled regions of China, comes forth a type of people who, because of their low standards of living and their bodily and nerve hardiness for performing their traditional labor, eliminates every competitor from the field, both at home and in the lands where they settle. Herein lies perhaps the greatest danger—the spread of this Chinese culture over the earth even more extensively than it is today!

Moreover, the lack of wood in China brings another problem with it—the impossibility of obtaining fuel for cooking and warmth.

No one needs prophetic gifts to realize that a human catastrophe is in preparation in those regions. This will actively affect the inhabitants of Europe, as well. Dr. Steiner once suggested that the student of human affairs would do well to investigate the influence of food as the driving unrest—producing motivating power behind the migrations of peoples. Races that had remained for long periods of time on the same soil have felt themselves becoming restless in the development of their common life, because of their one-sided nourishment, and have sought for a balance.

In short, what is the situation of Europe that occupies a position between the two extremes of the East and West? Europe has a temperate climate and a healthy division between woods and fields, lakes and plains, hills and moorlands. In the 1800s, there was still a family farm culture firmly based on the soil and those who tilled the earth based on traditional wisdom. However, engineering methods and science, along with economic difficulties, have already taken hold of agriculture, so that the once-healthy structure of the soil is beginning to fail. The place of the family farmer is being taken largely by agriculturists with no tradition, or "tiller mechanics" who seem no longer to have any relationship to the problem of soil, which is being attacked by a worldwide sickness.

Those who are candid and honest in regard to the facts of modern agriculture realize the existence of this situation, though they might not want to face it. Farmers know it through their feelings. It is known to research scientists, whose desks are piled with reports on unresolved issues of soil fertility, problems of insect pests, and so on. A leading agriculturist said recently: *"We have been exerting ourselves for the past two decades to halt this process, but our efforts have been fruitless. What is the solution? Where can we find the answer? The disease caused by decreasing natural soil fertility is a global sickness."*

CHAPTER 3

The Farm and Its Broader Connections

The farmer has not only to deal with his soil and with his seed; he is connected with an encompassing life process in his wider surroundings. One simple fact may be mentioned in this connection. Although the farmer's land is spatially limited, the plants on it draw only from two to five percent of their nourishment from the mineral substances in the actual soil itself. The remainder—water and carbonic acid—originates from the air. But through its relation to the atmosphere and its currents and movements, the plant world is connected to the earth as a whole; therefore, its sustenance can come from beyond the limits of individual continents.

As a result of the movements of the air, due to wind and to high- and low-pressure areas, it may happen in Europe, for example, that the air comes at one time from the middle of the Atlantic Ocean, at another from Greenland, or again, from Northeast Russia. A cold wave, an inrush of cold air from Greenland, has quite a different character, for instance, from one coming from Northeast Russia or in the United States from Canada. The former brings a moist cold, the latter a biting cold, but dry air. Furthermore, numerous observations have shown that the whole physiological state of an organism is dependent on these atmospheric influences. In the human body, blood pressure, propensities to every possible sort of organic disturbance, attacks of grippe, and so on parallel variations in atmospheric conditions. But such influences, though rarely observed, also play a role in the plant kingdom. Thus, for example, mildew might appear on spinach overnight, brought on by changes in weather. Moist air from the Atlantic, which has a tendency to form mists and fog, has a fostering influence on the prevalence and spread of fungal diseases, exemplified by mildew, that suddenly appear with such a change in the weather.

Therefore, in regard to climatic conditions, we see agriculture linked to the larger earth sphere. This applies to wooded areas, fields, prairies, bodies of water, and moors. Together, in a larger sense, they form a self-contained organism, whose parts (woodland, field, and so on) join together as the members of a living whole—the *total fertility* or *total life capacity*—of a country or an entire continent. All these members and organs are interconnected, each belonging to and depending on the other. Whenever such an organic unity is altered at some point, it also means a change in all the other organs. A level of maximum effectiveness depends on all of these factors working together. Striving to attain such effectiveness should be the task and goal of every agricultural measure.

The foundational conditions of life in Central Europe allow the use of land to its maximum capacity. Favorable factors consist of the climatic situation and the distribution of hilly land and mountains, lakes, moors, forests, and arable areas. Even the North German Plain is broken up at just the right point by mountain ranges. However, a diminution of the currently forested area by even 15 percent would lower the biological critical point of effectiveness and bring about new climatic conditions that, over the next few centuries, would produce effects tending toward the development of a climate similar to that of the Russian steppes.*

---

* A recent address to the Austrian Lumber Commission pointed out that the world's stocks of the Conifer—i.e., the stocks now available for commercial purposes—will be exhausted in about 40 years. This is true if deforestation continues at the present rate. Because the use of wood for all sorts of industrial purposes (paper, artificial silk, cellulose, plastics, and so on) is increasing enormously, we are faced with the danger of a far more rapid disappearance of the wood supply. The problem of increased formation of steppes is not only a well-known danger for the US, but the subject was recently discussed seriously in Europe because of the migration of Russian

Considered from the aspect of a general land hygiene, of soil use and the maintenance of soil moisture, moorlands and swamps perhaps may seem unnecessary, even harmful. But do we realize that a boggy moor, under certain circumstances, can be the source of the moisture so essential for forming the dew in the surrounding plain in dry times? The drying up of such marshes means a lessening of the dew formation just at the time when it is most needed. Drainage operations, in fact, can add a few extra square miles of land for cultivation. But by this process there is the risk of lowering the value of a much larger surrounding area that may be in the highest state of cultivation. Naturally, in these remarks nothing is said against an intelligent handling of swampy regions for the purpose of fighting malaria and other diseases. The essential point is that in every case an adequate amount of water should be allowed to remain for the purpose of evaporation. But this water can be kept flowing by proper regulation and need not stagnate. There is another, seemingly unimportant sort of local change, the consequences of which I have been able to observe personally. Numerous elms were planted during the 1800s in Holland, and a monoculture of this beautiful, stately tree covers the flatlands along the seacoast. This growth is not really a woodland, but merely an interruption of the plain by groups and lines of trees. They create a lovely landscape, but also form a valuable screen against the wind that at times blows in from the sea with great force, chilling and drying the soil.

In recent years, however, Dutch elm disease has largely wiped out this fine stand of trees. Now the winds blow unhindered over broad stretches of the plain. My own practical experience in this connection demonstrates that it has now become impossible to replant a similar line of trees, because between the fallen row of elms and the shore dunes, about two miles beyond, other trees that had grown in lines and clumps have also fallen. The force of the wind can now

steppe plants into Czechoslovakia and even into the most fertile lands of southern Germany.

come unhindered from the sea, so that a second growth would be unable to withstand it alone. The result has been that new young roots were torn to pieces and dried out. Even where a tree here and there did manage to reach a height of more than nineteen or twenty feet, gales eventually broke it down—and this in a locality where a mighty row of old elms once stood.

Tree planting in this locality would be possible only gradually and with care. Beginning at the coastline, protective vegetation would have to be built up. Behind the shelter of dunes and the bushes growing on them, young shoots might get a start. After reaching a certain stage of growth, they would in turn serve as protection for another line of trees planted farther from the shore. The accomplishment of such a plan, however, would require the cooperation of neighboring landowners. Unfortunately, an intelligent vision for this is lacking. Consequently, the topsoil is drying out through the strong force of the wind; the ground is becoming chilled, and the water table is being lowered.

A landscape architect with whom I discussed this problem felt that what had happened was actually desirable—the flatness of Holland was now coming to expression. He believed there was an aesthetic advantage in lieu of hedges and trees. Poor Holland; its very name (*Houtland* = woodland) indicates the existence of a very different landscape in the past! Surely, fruitfulness of the land is most important, and hedges are the most essential factor in this sense. Hedges grown to six or seven feet can block wind from the ground for a distance of nearly 500 feet, raising the ground temperature one or two degrees for a distance of 300 feet.

Contributions to a farmer's soil arrive from a distance as dust. There are many natural sources of dust, one of which is volcanic eruptions. The famous eruption of Krakatao ejected dust into the air that traveled around the world for several years as clouds and could, even after thirty years, be detected in the atmosphere. The ashes of Vesuvius that buried Pompeii to a depth of several feet are a classic example. They produced especially fertile soil. Fine ash from that eruption was carried as far as Anatolia. More Recently, a volcanic eruption in South America emitted ash that perceptibly darkened the atmosphere and reached as far as Europe. It has been calculated that the fine, invisible dust reaching Europe from the Sahara amounts in a hundred years to a layer 5 millimeters thick. In the American Midwest during the past 100 years, about an inch of soil has been blown from the western prairies into the Mississippi basin. Annually recurring great dust storms deposited 850 million tons of dust per year in the basin, from a distance of more than 1,250 miles, even before the recent disastrous [1930s] Dust Bowl.

The Sahara dust storm of February 1901, which affected an area of about 170,000 square miles, brought about two million tons of material to Europe, and distributed 1,600,000 tons in Africa. A year later, another dust storm from the Canary Islands brought about 10 million tons to England and Western Europe. Such dust is either brought down by rain or, in passing, it adheres to moist ground and vegetation. In China, windblown deposits reach a depth of several hundred feet. Such deposits aid in the production of fertile soils. Curiously, the dust deposits of the American Midwest have a composition similar to loess (flood deposits of loam).* Prof. Passarge tells us that the black earth of southern Russia is the best and most typical development of loess, which favors the development of steppe flora and, because of its porous quality, is easily impregnated with humus materials.** It would seem that the black earths of Morocco are also a loess formation.

According to T. Fischer, southerly winds from the steppes west of the Atlas Mountains bring dust, which is retained by vegetation and precipitated especially by heavy dew at night. Dust deposits and decaying vegetation are thus the origin of the extremely fertile black soil.

---

*   Loosely compacted, yellowish-gray deposits of windblown sediment.

**   Passarge, *Die Wirkung des Windes. Handbuch*, vol. 1.

With a content of from 25 to 34% organic matter and a remarkably high potash content of 1 to 4%, and even 6.9% in volcanic ash, these deposits from the atmosphere constitute a well-balanced fertilizer. Volcanic ash is the cause of the fertility of the plantation lands of Central America and the East Indies. Chinese farmers welcome dust-laden winds in the Yangtze basin as suppliers of fertility, as did farmers in the prairies and South African veldt, until the catastrophes of recent years. Indicating the importance of the quality of annual deposits of atmospheric dust, Dr. Treitz summarizes his conclusions:

1. Dust deposits restore the exhausted bases and soluble salts (calcium, iron, potassium, and so on) in the soil.
2. Inoculating substances for stimulating life in the soil are carried in the atmospheric dust.*

Braun, Blanquet, and Jenny have shown that the soil of Alpine meadows is formed largely by dust particles from neighboring mountains; deposits of 2.8 to 3.7 pounds per square yard each year were measured. Discussing the nature of peat ash—which, in addition to considerable water content, also contains mineral ingredients, especially argillaceous sand, magnesia, gypsum, oxide of iron, in addition to some alkalis, phosphoric acid and chlorine—those same authorities assert that at least a portion of those substances could have been brought into the pure peat moss only by the wind. The quantity is considerable, since dry peat moss contains an average of 10% mineral substance.

Heavy deposits are also made through the medium of the air in neighboring factory towns. It is calculated that 224 to 296 pounds of soot and dust per acre each month fall to the ground in industrial areas. A third of this deposit consists of soot and sulfuric acid. Additional fertilizing substances are also present in the rainwater

---

*   From the report contributed by W. Meigen, *Material aus der Atmosphäre, Handbuch der Bodenlehre*, vol. 1.

received by the earth. For example, rainwater contains carbonic acid (H_2CO_3), an important fertilizing element, of which more than a million tons fall annually in rainwater in Germany alone. Along the seacoast are salts brought by wind and rain. Then small but active quantities of iodine (0.1 to 0.2 mm^3) are present in the air. Also present is chlorine. Along the coast of England, one quart of rainwater carries at least 55 milligrams of chlorine, and 2.2 milligrams are deposited more inland. During storms in Holland, 350 to 500 milligrams per liter of rainwater were measured. This means that in those areas we have chloride fertilizing from the air, ranging from 1 to 13 pounds per acre. In many tropical regions such as Ceylon, this figure rises to 30, 50, and even 150 pounds per acre, owing to the strong evaporation of the seawater.

It is also surprising to find that other substances important to plant nutrition are present in measurable quantities in rainwater. There is a special relationship between rain and nitric acid (HNO_3), along with other nitrogen compounds, and between phosphoric acid (H_3PO_4) and snow. This is important for practical agriculturists. In their desire to conserve soil nutrients and to free those bound up in it, they often resort to one-sided chemical fertilizing, not realizing the assistance nature also offers. Nitrous compounds rain down on England, for instance, at a rate of 3½ pounds per acre, and in northern France at a rate of 9 pounds per acre each year. Measurements of sulfuric acid yield the following figures per acre in Germany: Giessen, 8 to 100 pounds; Cologne, 200 to 340 pounds; Duisburg, 200 to 600 pounds per year—certainly considerable fertilization.

Another source of material from the atmosphere should be mentioned—so-called cosmic dust and meteorites. A significant quantity of such cosmic substances reaches the earth throughout the year. A continuous change of substance takes place between the earth and cosmic space. In the case of meteors, many are concretely perceptible, since individual specimens often weigh many tons. To this we must add shattered, dust-like particles from shooting stars. The quantity of such cosmic dust varies from 10 thousand to a million tons per year. It even causes the formation of certain red deep-sea sediments.

The mention of these facts is not intended to imply that everything needed by farmers comes from the air and that manuring is therefore unimportant. Good manuring remains the basis of all agriculture. These facts are cited only to point out that there are often active factors in addition to those involved in the nutrition equation. There are more processes interacting between soil and plant in nature than there are in a test tube. Extending our knowledge in this way is vitally needed. We should learn to understand and be able to reckon with the biological processes in nature.

Mineral substances are collected by the plants themselves and thus brought into the soil. This is discussed later in connection with plant nutrition.

CHAPTER 4

Soil as a Living Organism:
The "Load Limit" in Agriculture

S oil contains dissolved and undissolved mineral ingredients, water, organic matter from living plants and also organic and inorganic substances that have originated from the decay of roots or entire plants. Furthermore, it contains such living organisms as bacteria, earthworms, insect grubs, and sometimes even higher animals, whose activity and decay contribute to the transformation of the soil, physically by breaking it up, and chemically by feeding, digesting, etc. The interaction of all these factors together with climate, daily and annual weather conditions, and the cultivation that the human being gives it, determine the fertility of the soil.

The purely mineral, solid rocks of high mountains are infertile. Once its surface has decomposed and become porous through heat, frost, and rain, it becomes possible for the beginnings of life to appear in it. Weathering gradually creates a crumbly top layer, or soil surface. The more intense and rapid the weathering process is, the more fertile this layer becomes. This cyclic weathering process repeats year after year and acts on recently disintegrated rock, as well as on every piece of soil already tilled; physical disintegration of soil takes place through frost and rain. During summer, through warmth, radiation, and life processes in the soil (bacteria, etc.), dissolution of a more chemical nature occurs. The weathering process can reach the point where various constituents separate, and everything soluble is carried away by water.

Sand and clay form the main mass of alluvial sands, in contrast to weathered soils that include all kinds of mineral substances such as silica, volcanic rock, chalky marl, and so on. Individual rocks yield substances susceptible to weathering. Thus, volcanic lands, because of the igneous process it has undergone, are constituted differently—for example, from northern mica schists. Sand and clay disintegrate rapidly with fertile results, solid rocks reluctantly, providing sparse soil. Limestone is especially resistant to weathering, and covers itself with only a thin layer of marl. In more southerly regions, chemical weathering, especially oxidation, is more intense under the influence of the sun's radiation.

There it is often a lack of water, which keeps the ground from becoming fertile. Thus, every soil, according to its geological ancestry and age, represents a particular condition of weathering. Organic processes cooperate; disintegrating plants and animals eventually furnish the humus. The quality and amount of the humus determine soil fertility. Not all organic soil constituents need to be humus, or even potential humus. Only about 40 percent of the organic constituents are humus or humus-like. The actual content of neutral colloidal humus substances is relatively low, even in soils rich in organic matter—usually less than 1% in highly cultivated agricultural land in a temperate zone.

Just as the inorganic ingredients of soils weather in different ways according to climate conditions and the sort of soil cultivation, likewise there is even a greater difference and wider variation in the organic content. The most important of the factors governing decomposition are an intense aeration of soil, a lengthy drought, heat, intense solar radiation, all of which break down the substance into ammonia or nitrogen and water. In other words, these things are lost as fertilizers for agriculture. A hard soil surface, moisture, cold, extensive cloudy weather, water accumulation in the soil, and insufficient drainage all bring about humus material rich in carbonic substance and poor in oxygen. Such soil gradually grows increasingly sour, and if air is kept away from it completely, it eventually turns to peat. Such soil is indeed rich in organic substance, but this material is sour and in a matted state, and hence not directly useful. The valuable neutral colloidal humus is missing. Organic substance of this kind is a dead mass insofar as agriculture is concerned.

It must be pointed out here that there is no exact chemical formula for humus. We know only that we have in humus a mixture of the products of disintegration, more or less rich in carbon, nitrogen, and oxygen. In humus, the important factor is its condition. The degree of organic disintegration, along with the state of the soil—chemical and physical—determines its value.

Soil bacteria, as well as earthworms and their digestive activities and decayed bodies are the main participants in forming humus. In rich and healthy soils, earthworms furnish as much as 600 pounds of humus per acre—considerable fertilization! This is even more valuable, insofar as the bacteria are gatherers of phosphoric acid, their bodies being rich in this substance. They also assist in freeing the phosphoric acid already present in the soil. Of course, phosphoric acid is generally present in only small amounts in an available free state. It is the task of these minute organisms to help quicken the soil by liberating its natural reserves, thus adding to the phosphoric acid in manure and compost. In addition to this (according to research by the Swiss chemist Paulus), the atmosphere itself holds finely diluted quantities of phosphoric acid. During winter, measurable quantities of it are deposited, especially in snow.

The process of nitrogen fixation by bacteria is well known. Might it not be possible that a similar process exists in the case of phosphoric acid? This question deserves investigation and research.

The author is acquainted with the studies of noted research institutions reporting that, on large test areas, no diminution of phosphoric acid has occurred, despite intensive cultivation of the land. Although no phosphoric acid was added to that provided by normal manuring of the land, it was observed that the phosphorus pentoxide (P_2O_5) content of the same soil was subject to fluctuations at different times of the year—indicating that we are dealing with organic processes in the microscopic world of soil.

The acidity degree of a given soil varies with the seasons, showing the minimum during autumn months and the maximum during cold winter months, with two additional secondary minima and maxima during the year. There is another variation, depending on the natural sour or neutral condition of the soil. Sour soils (in the Northern Hemisphere) have a minimum in June and a maximum in March; neutral soils, the minimum in May and October and the maximum in February and August. It will thus be necessary in the future to know exactly when soil analyses are made, since they can be compared only with other tests made at a similar time of year. Periodic variations with two minima and maxima have also been detected in the humus and nitrogen content of soil; a cyclic curve for phosphoric acid and its solubility in forest soil has also been observed, showing a minimum during summer and maximum solubility in autumn and winter. According to Prof. Fehér of Sopron, phosphoric acid analyses with no indication of the time of year have no practical value in view of the seasonal variations of its soluble or insoluble compounds.

Bacterial activity in soil shows a maximum, respectively, in late spring, early summer, and early autumn; naturally, there is a period of rest during the frost season, as well as a period of rest in midsummer. From this, relatively new perspectives arise for agriculture and forestry that should place these rhythms in conscious connection with the whole biological treatment of soil and manure. Here a new field of research and practice is opened up.

In addition to this, we must consider the preference of certain plants for certain substances during certain seasons. In the case of some leafy trees, a rhythmic change of phosphorus and nitrogen was observed in June. The trees' outer rings show a maximum content of magnesium in April, a smaller maximum in November, and a minimum in August. The inner rings, to the contrary, show a maximum in August. Furthermore, seasonal deviations in the iron and calcium content, chiefly of perennial plants, forest trees, and so on are well known. Hence the problem of phosphoric acid appears not to be a part of inorganic soil science, but something that is essentially connected with the whole humus problem, and therefore worth referring to at this point.

In temperate zones, the most important makers of humus in cultivated land are earthworms, which digest organic refuse combined with the mineral soil components, and then excrete humus. Earthworms also help with soil drainage and aeration by making small holes and passageways. In virgin soil, humus gives natural fertility and a light, crumbly structure, even to heavy soils; light soils are also protected against drying out and erosion through the capacity to hold water derived from the humus substances excreted by earthworms. It is not a surprise that Charles Darwin, as early as the nineteenth century, dedicated a whole book to the fertilizing activity of earthworms.* He realized that 10 tons of earth per acre annually are worked over by earthworms, suggesting that without these worms there would be no soil for agriculture. Since then, there have been numerous studies of earthworms. It is estimated that the soil of a meadow in good condition contains 200 to 300 pounds of worms per acre. For farmers, the presence of a numerous earthworms in the soil is a visible sign of natural biological activity. The use of a microscope is impracticable for farmers, so they must use this natural barometer of fertility.

Every measure that disturbs soil life and drives away earthworms and bacteria makes the soil more lifeless and less capable of supporting plant life. In this connection we see the dangers of one-sided fertilizing, especially when one uses strong doses of chemical fertilizers containing soluble salts like potassium or ammonium sulphate, or

---

*   *The Formation of Vegetable Mould through the Action of Worms, with Observations on Their Habits,* London, 1882.

highly corrosive substances such as nitrophosphates (usually under a trade name) or poisonous sprays such as arsenic or lead preparations, all of which injure and destroy the microorganic world. Soils intensively treated with chemical fertilizers or orchards sprayed over a long time with chemicals no longer contain biological activity. The author has seen vineyards treated for years with copper and lime solutions; they are absolutely devoid of earthworms—that is, devoid of creators of new humus.

Once the humus content is exhausted, new conditions arise in the soil. The mineralized structure comes to resemble the laboratory conditions of a purely chemical relationship between soil and plant. This law is then completely valid: *What is taken from the soil in mineral substances by harvest must be returned to it.* This statement is true insofar as laboratory research is concerned. This rule was discovered and established within a closed system of containers in a laboratory.

This same rule would indeed hold true in nature, as well, if nature were composed only of lifeless mineral components. Through our modern intensive methods—especially in our use of strong or exclusively chemical fertilizers—we have created conditions so that the purely physicochemical qualities of soil are made to predominate, while organic activities are eclipsed. Mineralizing the soil becomes visible with the disappearance of earthworms and the formation of a surface crust that can be lifted off like a shell in dry weather. Farmers should consider the latter phenomenon as a "storm warning" from the soil's "barometer."

When the winter furrow, weathered by frost and rain, shows a porous, uneven, and irregular surface on damp days in spring, or when harrowed land remains porous, light, and crumbly after a rain, or when in summer, after a long dry period, it shows little or only a thin skin with the earth dried in fine cracks, then biological activity is present. We can call soils of this type *elastic,* not only because of the springy, elastic feeling one has when walking over them. This

term *elastic* is even more descriptive of the relatively resistant powers of such soils against all sorts of harmful influences—for instance,
mistakes in crop rotation or in tilling or harrowing when the soil is
too wet. Above all it is elastic in its retentive capacities with respect
to water. Water, which is not only important for plant nourishment,
but also makes up some 40 to 80 percent or more of plant substances,
should under all circumstances be retained by the soil. An important sign of a living soil is its capacity for absorbing water. When the
soil is alive, a sudden summer shower is immediately absorbed and
disappears as if into a sponge. When puddles remain on the ground
in wet weather, despite the fact that drainage is otherwise functioning well, it is a sign that conditions are not good.

An impressive example of such conditions was seen by the author
on a visit to the Agricultural College at Missouri State University.
A large experimental plot had been divided into two parts. During thirty years, one part was left untouched as virgin prairie; the
other was tilled for three decades in the manner of modern, intensive cultivation. As a result, the virgin prairie, though a heavy, dark
brown clay, was porous, so that a stick could be driven to a depth
of more than 18 inches. Rainwater disappeared immediately. Even
today, four years after that visit, the following experiment could
still be made on samples of this soil taken away and preserved in the
author's laboratories.

If some of the untilled soil is placed in a test tube and covered
with water, the water is immediately absorbed and a damp mass is
the result. After drying it out, the test can be repeated at will.

On the tilled half of the field in Missouri, the surface of the earth
had become increasingly hardened, and a stick would penetrate only
a few inches into the ground. Rainwater stood in the furrows, and it
ran off wherever there was a slope, making channels—the beginning
of dreaded erosion. With such mineralized soils, the danger of erosion
from rains, accompanied by a drying process, is very great. Every
good farmer living in hilly country and in hot climates knows this.

Mineralization of soil can be caused through destruction of the physical and organic structure. We speak further on about the influence of cultivation. The effect of using chemical fertilizers on plant growth is well known. They increase the yield, producing thick, heavy plants, especially when nitrogen-rich fertilizers are used. This effect has made farmers especially fond of chemical fertilizers; they yield larger harvests. Scientists even tell farmers that this supplies otherwise deficient elements in the soil.

However, two things may be observed by practical agriculturists who live and work close to the soil:

1. To maintain the same yield from year to year, we must frequently increase the annual quantity of chemical fertilizer.
2. The soil structure changes toward the hardening and encrusting process described here.

Why do agricultural schools and researchers remain silent about these phenomena—so familiar to a significant number of practical agriculturists? Much is said about the value of artificial fertilizers for increasing yields, *but very little about alteration in the soil.* Perhaps this is because the soil used for parallel tests in experiments is already in such a condition that this alteration in soil structure is no longer obvious.

Occasionally, in scientific literature we find a note to the effect that chemical fertilizers must be given in the correct proportions according to soil analyses. However, in most cases a field is not uniform in soil content; theoretically, therefore, we would have to use a different formula for each of its variations. That would be especially difficult in soils produced by weathering. As experts in the science of fertilizing readily admit, this differentiation is indeed theoretically necessary, but impossible in practice. Thus, they take an averaged formula and accept unbalanced fertilizing as inevitable in practice. Another point to be considered here is that not all farm managers are in the position to have an adviser standing at the ready and who is experienced in the problem of *their particular soil.*

In regions where there is enough moisture in the form of atmospheric humidity, subsurface water, and rain, fewer difficulties are encountered initially in terms of hardening and the alteration of soil structure. The reason for this is that the normal admissible balance of mineral salts is restored through the balancing, distributing, and dissolving effect of the water. In Holland, for example, the negative effects of artificial fertilizers appear much more slowly, owing to the high level and the constant circulation of the water table there, which provides a natural equalization.

But what happens during a dry summer or in an arid climate? In addition to the encrusting of the surface, another phenomenon might be observed. Small, less fertile patches will suddenly appear in a field. They grow from year to year, and in dry years the sterile spots will increase. We can ascribe these to so-called free-acid products—that is, insoluble or nearly insoluble silicates formed from the soil silicates in the interchange of salts. These insoluble silicates must be considered lost to the substance cycle of agriculture, and they can be restored only after a lengthy organic cure.

To explain the foregoing, we need to remember that, according to physicochemical laws, a state of equilibrium exists among the various salts in a solution. If one adds an easily soluble salt to a mixture of salts already in solution, the less easily soluble can thereby be forced out by precipitation (Gibbs free energy). This process is widely used in the chemical industry—in the precipitation of organic dyes, for example, by means of sodium chloride, or ordinary table salt. In soil, there is a mixture of salts; silicates that dissolve with difficulty; salts and silicates easily washed out by weathering; and easily soluble salts, especially potassium salts; and salts arising from organic ingredients. When easily soluble fertilizer salts are added, the balance turns in their favor against the heavy silicates. In other words, the weathering and disintegration of the ingredients of the soil are slowed down, and in unfavorable cases are made impossible. Plants are then dependent on the soluble fertilizer salts, and they react clearly to their presence or deficiency. Under such conditions, soil chemists are indeed correct. The problem of developing a buffer capacity belongs among these questions.

Soil biologists start from other premises. For them, the earth represents a natural reservoir, and except where pure sand or solid rock is present without water, a more or less complicated mixture of salts, silicates, zeolites, aluminates, and so on will always be available. These mixtures represent the natural basis for the substance cycle when it is made available. *We should not precipitate the salts in our soil,* but instead open the soil and enliven it. Organic processes that take place in the soil due to light, air, weather, microscopic organisms, and humus are activities that help its "demineralization," on the one hand, and foster its "becoming organic," on the other. By doing so, they open up extensive reserves. As a result of a single year's harvest, when the soil substances are used up, new substances are freed again through the natural action of weathering in a living, vital soil. Tillage and the eroding effect of wind and water liberate new soil ingredients every year. These need only to be enriched—"organized" in the most literal sense of the word.

In Central Europe, we estimate a yearly loss of 1 millimeter from the surface of a field through wind, rain, and tillage, whereas in alluvial lands there is a constant addition of material. In the Nile Delta, for instance, this is helped by wonderful natural irrigation.

Yearly plowing and harrowing of a field, accompanied by the loss of a millimeter from the surface, has the effect that each year the depth of the cultivated portion of the soil moves 1 millimeter lower. In this new millimeter, a reserve mass of mineral ingredients is uncovered. The important thing is that this reserve should be liberated and added the cycle of agriculture. With a yearly deepening of only a millimeter of soil, about 600 tons per acre of new soil are brought into action; this constitutes, even in very poor instances, a valuable reserve. This soil includes potassium and above all phosphoric acid that in normal soil is transformed by microorganisms, but that, by means of various salts such as calcium, is in danger of being fixed in insoluble forms from which no good can come.

In studying such relationships, we see an important difference between conditions in a research laboratory and those in nature. If we also count on accidental sources of mineral substances during the course of the year—especially those brought by wind and rain—we can see how little a cultivated field constitutes a closed system of substances. We need only be careful in our methods of cultivation to preserve the living elements of that nature and not allow them to be precipitated. Among the occasional sources of additional substances, we should include the salt content of rain and snow, as well as the movement of dust from both organic and inorganic origins.

By observing the various colors of winter furrows at the close of several winters, we have a good example of how differently the weathering processes work on the decomposition and structure of the soil in different years. When we take all this into consideration and with it the fact that plant life, by having its roots in the earth, sets up a whole series of processes and transformations in the soil, it becomes clear that *cultivated fields are living organisms—living entities in the totality of their processes.*

The "Load Limit" in Agriculture

Scientific farmers today talk continuously about high production, which clearly depends on the overall capacity of one's farm. Capacity depends, in turn, on a variety of factors related to life process, as discussed. Thus, the first task is to carry on the work according to complete knowledge of the nature of the life process, just as in other areas engineers base their work on a careful study of the strength of their materials.

Why are we always talking about high production—*high production!*—while not thinking like construction engineers, which is based on real knowledge and an understanding of the "materials" to be used? Corresponding to the inorganic "load limit," there is equally a limit of inner capacity—a biological "elastic limit," or biological "resistance point"—for the "living material," or for the organism in the broadest sense of the word.* If the limits of these expansion values are overstepped, harm will come to the life phenomenon, just as the collapse of a mechanical structure occurs if the resistance point is overstepped.

A cultivated field does not have unlimited capacity for increasing productivity. Its productive capacity is not directly proportionate to the amount of fertilizer applied. A cultivated field is also a biological organism, and as such it is subject to the laws that govern the organic. It has its critical point of inner effective power. This is defined as the product of numerous varied factors, among which are mineral ingredients, physical structure, the presence of organic substances and their conditions (humus, acidity, and so on), the climate, methods of cultivation, the variety of plant growth, plant-root activity, ground cover and cover crops versus erosion, the possibility of weathering processes, proximity or the absence of woodland, and so on. All these factors together determine the capacity for biological activity. If any one of these factors is disproportionately active, a disturbance arises and, in time weakens the entire system. By strengthening fertilizing, production is increased. However, if this is overdone (in a one-sided way), the soil as an organic structure falls apart. The basic principle underlying healthy, long-term agricultural production is, therefore, knowing the "resistance point" of a specific soil. This alone gives assurance of lastingly healthy conditions, making productive capacity certain.

---

* The terms *load limit* and *elastic limit* are used in the sense of mathematical–physical observations of nature.

As a living entity, soil capacities have a certain limit. This is also true of a machine, which ceases to operate if it is overloaded; it breaks down and becomes unusable, but it does not restore itself. A living entity can bear a certain amount of overburdening. For a short time, more can be demanded of a draft animal than is good for it to give, but it will recover its strength in time. Only through long and continual misuse is it weakened and its offspring made less valuable. Such weakness, however, is not immediately noticed. When it finally shows, it is often already too late to correct the error. It is easier for a keen observer to prevent sickness than to heal an organism already attacked and partly or wholly injured. This is also true of soil; it has a natural standard for living. To recognize and guard that is the greatest art of farmers. If a certain field is expected to produce beyond its capacity, it can (especially through increased fertilizing) be whipped into being overloaded for a while. Then troubles begin, however, and the longer this continues the more difficult it is to eradicate the problem or cure the organism.

It is of the greatest importance for practical farmers to understand the natural load capacity of their soil correctly. Only when they actually have the same regard for their soil that they have for a horse—the welfare of which one guards daily—can farmers expect to get performance from the soil that is commensurate, year after year, with its capacities without harming it. Everything they do in connection with their soil must be regarded from this perspective—plowing, harrowing, seed choices, and manuring.

In terms of the manifestations of their life, neither soil nor cattle nor human beings is a mathematical problem. Every jockey knows that the performance of a horse—its pace, jump, and endurance—does not depend entirely on feeding, which obviously creates the physical basis, but the horse's protein or lime content does not transform itself proportionately into jumping ability. Nor are heavily fed horses the ones with endurance. Riders know there is much besides feeding that takes place between them and their horse, which has an effect on evoking its best performance. Unfortunately, farmers do not regard their cows in the same way. Cows have already become largely the subject of arithmetical calculation. In one end, x number of pounds of proteins and salts are stuffed, so that x number of gallons of milk will flow from the other. Cows cooperate for a while, but then we are surprised by weak calves, streptococcic diseases, abortions, and so on—phenomena that arise "inexplicably!"

Cows, through intensive feeding, can be brought to maximum yields of milk. This is the result of a one-sided performance on the part of the organism. Milk production is a part of the animal's sexual activity. Intensification on the one side means diminishment on the other—in this case, a weakening of the reproductive organs. We might get higher milk production, but as a result we also have to deal with all the familiar difficulties of breeding and raising cattle, such as contagious abortion (Bang's disease), delayed and difficult birthing, sterility, mastitis, and other streptococcic infections. The "expansion limit" of the productive capacity of the cow is a product of breed, constitution, size, quality and quantity of feeding, individual capacity for the utilization of food, pasturing or stable feeding, local terrain, climate, ability of the breeder, etc.

We have become even more accustomed to regard the soil as a real equation of nutritive values. Such an equation would, as a matter of fact, be correct, if we included *all* the factors. But the following is an example of one-sidedness and an improper equation: soil plus additional fertilizers equals soil plus yield.

The proper equation, however, from the point of view of life, should be as we see in the chart on the next page. Consideration, or neglect of any one of those factors is just as important as the whole fertilizer equation.

What then can farmers do to be fair in all these, while maintaining their living organism—*the farm*—at a high level?

Natural fertility plus production capacity equals the sum of biological functioning of

| soil |
| manuring |
| humus |
| tillage |
| rotation |
| climate |
| weather conditions |
| quality of seed |
| weed growth |
| several environmental factors |

Chapter 5

Manure and Compost Treatment

The first problem facing farmers is the proper use of manure and the correct treatment of compost. Contrary to popular belief, *we feed the soil by manuring;* we do not feed the plants. The vital activity of the soil must be maintained, and nature takes care of this by developing humus through the activities of bacteria, earthworms, roots that break up the soil, and weathering. A farmer's primary task is to aid these natural *organic* processes in the soil. When manure is applied, it must enter the soil in such a condition that it contributes to this work of nature, and it is least able to do this if it is raw and fresh. Unrotted manure, in its process of decomposition, actually *feeds* on the soil for a certain length of time. This happens because unrotted manure requires biological activity and energy to decompose. Especially objectionable is the absorption of the decaying products of half-rotted albumin that might be taken up directly by the plant roots, which can have a disturbing effect on plant as well as human health. We can detect the sort of manurial treatment that a plant has received in the familiar smell of cooking cauliflower for example. *In the kitchen, we can actually smell the pig manure, sewage, and so on that have been worked into the garden soil.*

The best form of organic fertilizer is humus. Unfortunately, it takes a considerable length of time for stable manure to decompose and become humus. During this rotting period, valuable substances are disintegrated and lost if leakage is not checked. Usually, stable manure is exposed to an endless series of losses. To begin with, there is a loss of nitrogen under the influence of certain bacteria and the weather. These especially rob the surface of the manure mass wherever too much air is allowed to penetrate. The usual method of throwing stable manure loosely out into the manure yard and exposing it to the sun and rain can lead to a loss of 50% and more of nutritive substances. Some of its nutrition is leached away by sunlight, and oxidized away and washed out by rain. If manure sits on a slope, one sees a brown stream running from it, which carries away the most important nutritive substances.

Those same farmers who have carefully calculated from fertilizer tables the amount of ammonium sulphate they need for a profitable harvest will, oddly enough, watch with the utmost tranquility the nitrogenous substances of the manure run off as a liquid down the road, fertilizing the pavement and drains or, at best, the weeds along the ditch. I once met a man who was exceptionally "practical" according to his own estimation. Below his manure pile was a small pond, into which brown liquid from his manure pile was running in numerous rivulets. *"Oh, that's all caught in the pond,"* he said. *"We clean it out every four years!"*

Then there is the opposite sort of manure pile, one that gathers all its liquid at its base. This slowly soaks the base and rises into the body of the pile. This condition entirely insulates from the air the portion standing in water or liquid manure. This prevents proper fermentation, with the result that, instead of a good fertilizer material, we have a black, odd-smelling mass—a substance that is turning into a kind of peat. The manure value of this is slight. Such a product is more like a wet loam or clay than manure. It smears the ground and is difficult to take up; even after weeks, the black lumps can be found in the soil, still barely unaltered. Processes have occurred in this case that go far beyond the goal of humus formation, in contrast to manure that is leached out by too much exposure to the air. Manure that is, on the other hand, too tightly packed, runs the danger of heating up too much and losing its best qualities.

In short, there are numerous wrong ways to treat manure that can result in a loss of half its original nutritive value. Having calmly thrown away much of the nutritive substances in their manure, farmers rush to make up their loss by using mineral fertilizers. Even so, they soon learn that these help only temporarily.

The first and foremost practical rule for the correct treatment of manure is to heap it up daily with proper care in the manure yard. It is best to start in one corner of the yard and build a firmly trodden, but not tamped-down, rectangle, 2 to 3 feet high, with a base area of 2 or 4 square yards. Next, another pile is placed alongside it, then a third, and so on. The first section may be covered temporarily with a few planks, so that a wheelbarrow can easily run over it to the next. It is always very important, however, to cover the manure well. It is only when the entry of air into the heap is lessened by proper covering that the action of the bacteria that rob the surface is arrested. The best covering for the piles is earth. With medium soil a four-inch covering is enough, if the soil is heavy less is needed. The leaching and drying out processes are prevented by covering. Only where no soil is available should the covering be of peat moss, planks, straw, or potato vine thatch. If potato vines are used, it must be realized that a certain amount of their own valuable substance will be lost as they rot on the surface.

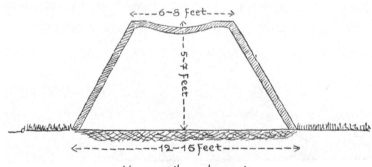

Manure pile and covering

The thickness of the earth layer, as indicated, depends much on the kind of soil used. However, the content of the pile should never be completely closed off from outside influences. Impermeable soils such as heavy loam and clay must be applied lightly, especially when these are used in a wet condition. When sand is used, care must be taken so that it is not blown away by the wind or in danger of sliding down the sides of steep piles.

The guiding principle in all of this is that, because of its bacterial content and its internal fermentation, a manure pile must be treated as a living organism. As such, it must have an outer boundary or skin to separate it from the surrounding environment; it needs to develop its "own life." The decomposition of manure should not be subject to a haphazard fermentation. The one aim of the manure pile is to produce humus, and the purpose of all useful organic decay is to produce a *neutral humus*. Manure brought to the soil in this state not only gives it the maximum in fertilizing value in terms of available nutrients, but it is also most helpful to the physical structure of the soil.

At this point, in addition to handling manure properly and carefully, the biodynamic method of soil treatment begins, as presented by Rudolf Steiner. Its effect on the soil is similar to making bread, in which water and flour are first mixed into a dough. This is left to stand for a certain length of time, exposed to air, so that any ("wild") yeast bacteria that might be floating about will settle into the mixture and, after several hours or days, cause fermentation. The bread baked from this dough is sour, bitter, and hard—and inedible. To produce a good, edible bread, bakers use a select strain of cultivated yeast, or perhaps a leaven, to obtain a quick and beneficial fermentation. Farmers typically treat their manure according to the first method—fermentation is left to chance. The proper course is to develop controlled fermentation that allows only a minimum loss of nutritive elements and leads to better humus formation. Thus, one follows a controlled method, not one left to chance.

Dr. Steiner has shown that we can gain such control by using certain plant preparations to induce the right kind of fermentation. This is done through the use of various plants that have also always been used as medicinal herbs, including chamomile, valerian, nettle, dandelion, horsetail, and so on. Such plants are first put through a long fermentation process themselves—buried at certain depths in the earth and in close contact with specific parts of an animal organism. The process can be described by saying: *Through a kind of hormonal influence, fermentation is guided in a definite direction.*

After several months, the plants are actually transformed into humus-like masses. If small amounts of these preparations are inserted into a carefully piled manure heap, the entire fermentation of the pile is given the proper tendency toward humus formation. The result is that, after a short time (generally two months), the dung is turned into a blackish-brown mass, rich in humus content.

Researches have shown that, during the rotting process, the bacterial content of such a manure pile is 10 times that of one not treated in this way. Especially noticeable is the presence of a large number of earthworms. Such piles continuously fill with earthworms, which, after their humus-forming activity is complete, die and provide an additional fertilizing substance through decomposition of their bodies. To get the results outlined here, we must also consider a number of other points.

Strawy manure contains lots of air and heats easily, especially when it includes horse manure. Wet, greasy manure becomes putrid. The kind of manure produced also depends largely on the feeding process in the stable. The best manure comes from feeding roughage—grass, hay, clover, pea vines, or other straw. It has the structure most favorable to fermentation, especially with the liberal use of straw for bedding the animals. Turnips, turnip leaves, and so on produce manure that is too wet. Concentrated feeding produces sticky, wet manure, which ferments slowly or not at all. The most unsatisfactory manure comes from animals fed mainly on

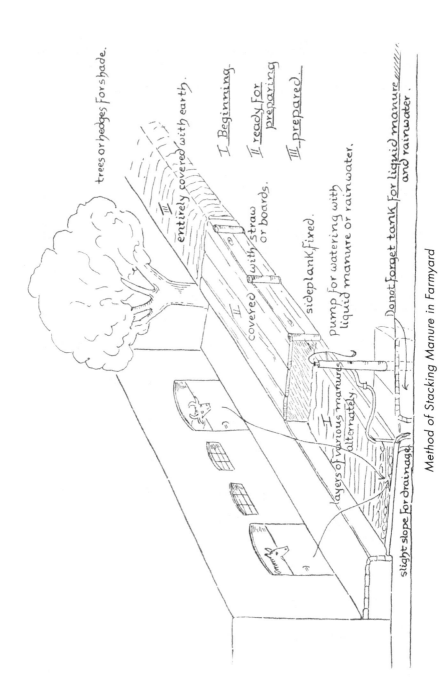

trees or hedges for shade.

entirely covered with earth.

I Beginning.

II ready for preparing.

III prepared.

III

covered with straw or boards.

sideplank fixed.

II

pump for watering with liquid manure or rainwater.

I

layers of various manures alternately.

Do not forget tank for liquid manure and rainwater.

slight slope for drainage.

Method of Stacking Manure in Farmyard

concentrates with little hay, as well as when only materials such as leaves, sawdust, and so on are used as bedding instead of straw.

Departure, even partial, from conditions that produce the best manure can be easily detected, and corresponding measures can be taken to correct them. Because the best preservation of fertilizer values comes from using mixed manures—horse manure being especially protective against bacterial denitrification. A mixed pile is recommended, but of course only when this is possible. Anyone who uses horse manure for hot beds must naturally store it separately. Otherwise, the best practice is to take cow manure—produced from hay and green feed and caught up in straw—and spread it carefully out over a small area in the manure yard. A thin layer of horse manure is then put directly over this base layer of cow dung. This combination remains lying for a number of hours to "steam out." Then, before the next layer of manure is added, the base layer is trodden down until it becomes somewhat firm. When the manure is strawy, this trampling down can be done more vigorously. If it is too sticky and wet, it must be less solidified. If the manure originates from concentrated feeding and has little straw mixed with it, it must lie exposed to "steam out" in the air a little longer to dry out somewhat. Whenever possible in this case, two piles should be started alongside one another, with additional piles made alternately.

The heaps can be covered with earth, boards, or straw once they have reached a height of 3 to 4 feet—strawy manure shrinks together more than other manures and can be piled higher than wet manure, which suffers from too much pressing and packs itself too solidly together. A new heap is then started, or additions are made to the old one. The sides can be made of movable pegged planks, which are removed when the new batch is set up alongside. When making long piles, the procedure is to cover 3 to 4 sections as soon as they are set up. Such complete sections must then have the biodynamic preparations inserted from all sides to obtain properly controlled fermentation from the very beginning.

A biodynamic manure heap

Cross section of a 3-to-4-month-old biodynamic manure heap; the straw has
completely disappeared. A beautiful, brown mass is now
available to spread over the field with a shovel.

If for any reason the manure is too wet (this arises from improper feeding or from too much liquid manure, lack of straw, or excessive rain), a provision for drainage must be made within the pile. The manure pile must never stand "with its feet in the water." If the manure is too wet it cakes firmly together and becomes greasy and unable to get sufficient air for fermentation. In such a case, when the addition of straw during the piling does not help sufficiently, the required and proper aeration can be obtained by drainage.

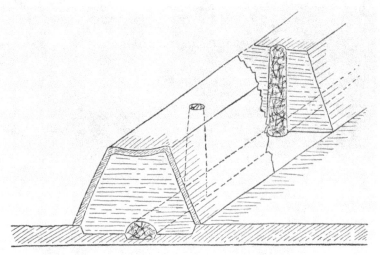

Drainage scheme in a manure pile

The necessary drainage can be effected by making a core of thorn brushwood, briar canes, or by using perforated drain pipes. It is advisable in very bad cases of dampness, or where the pile easily gets overheated, to add "ventilators." These can be opened or closed as needed. The liquid running out of the drains should be gathered in a tank located at the end of the manure yard.

A manure pile that is too dry requires watering. Dry manure gets hot very easily, and when there is no moisture at all the fermentation induced is disintegrating and destructive. The pile remains unchanged or, when some chance wetting occurs, mold might form.

This whitish-gray growth, as well as the presence of wood lice in manure and compost, is always the sign that treatment of the manure is too dry.

The correct moisture condition for manure is that of a damp sponge; no liquid should flow out of it, nor should it be stiff and dry. For good fermentation, this state should be maintained under all conditions. The best means of adding moisture is through pipes inserted in the upper part of the pile. Under extremely dry conditions, this becomes a necessity. Ordinarily, however, a trough-like depression on top of the pile is sufficient. Water or liquid manure is poured into this trough, and it slowly seeps into the pile. The best liquid for such use is what drains from the manure itself, the next best is liquid manure. Pumping natural liquid manure out of the reservoir tank or pit and onto the pile once a week to moisten it offers a special advantage. In this way, liquid manure is absorbed by the organic matter; it does not become putrid but is immediately drawn into the general process of fermentation. In this form, the liquid manure also has an inhibiting effect on the activity of nitrogen-destroying bacteria and thus preserves the nitrogen content of the manure. Its sharp, rank effect on plant growth is avoided in this way.

In many regions, especially the hilly areas of Central Europe, it is customary to carry the fertilizing liquid manure directly out to the pastures and hay fields. A clever farmer, whose field is situated high on a hill with the pastures lower on the hillside, once built a pipe system in such a way that he simply had to open a cock at the liquid manure pit to accomplish his fertilizing. The harmful effects of such one-sided fertilizing with liquid manure are well known. The reaction in the soil becomes increasingly acidic, characteristic pasture weeds appear, and the growth of clover gradually stops completely. Moreover, the high solubility of potash present in the liquid manure leads to a high percentage of this element in the fodder. This (as proved by the Swiss research scientist Friedrich von Grünigen), has in turn a baneful influence on the health of the animals. All such bad results can be avoided if the procedure previously suggested is followed.*

In general, manure that has been carefully piled up and given the biodynamic preparations ripens in two months and is transformed into a humus-like mass ready for use. If dryness or too much moisture or an excess of one type of the manure (too much horse in proportion to cow, for example) have brought disturbances into the fermentation, such as putrefaction, moldiness, or too much heat, it is advisable to turn the pile. This can be done when necessary after the second month. If the fermentation has already transformed the pile into an odorless mass after two months, turning is unnecessary. If while turning, too much dryness is observed together with too high a temperature (anything over 150°F. is considered harmful), then the mass must be watered thoroughly. Turning on rainy days is a good practice. Weeds should not be allowed to grow on manure or compost piles. A growth of grass on the piles is also harmful, since it cuts off air with its thick root system and checks fermentation.

On the other hand, in places without shade, the pile is exposed to too much direct sun. It then pays to shade it with straw mats, reeds, and so on. If necessary, however, to plant something on the pile for the shade, lupine, or vetch are suitable, or as seen more frequently, cucumbers or squash. The latter are planted in the ground around the outer edge of the pile, and the vines are trained up over it. A small hedge or trees should, correctly speaking, be planted around the site of the piles. In dry summers a variation in the speed of the rotting on the north and south sides of the piles has been noted. The shady side ferments somewhat more rapidly.

What has been said in a general way here about handling manure is equally valid for the preparation of compost. Compost

---

*   F. von Grünigen, "Die physiologische Bedeutung des Mineralstoffgehalts im Wiesenfutter mit besonderer Berücksichtigung des Kalis"; "Mitteilungen aus dem Gebiete der Lebensmitteluntersuchungen," in *Hygiene,* vol. 26, no. 3/4, Bern, 1935.

is a mixture of earth and all sorts of organic refuse that rots without having gone through an animal organism. It therefore lacks the presence of animal hormones, which, even in infinitesimal amounts, foster plant growth.

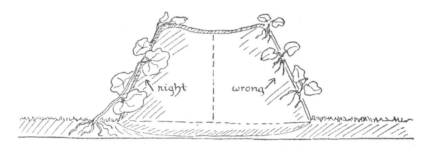

Right and wrong way of growing plants on a compost heap

Direct contact with the earth is necessary for compost and manure. Innumerable microorganisms, which until now have not been isolated in pure cultures, are present in the earth, and we are thus obliged to use earth as the source of those organisms. This is why it is necessary to build compost and manure heaps directly on bare ground and to avoid anything that causes a separation, such as concrete floors and so on. In this way, microorganisms and earthworms have free access to the heap. Even grass and turf constitute some separation, so it is better to remove the turf from the site of the pile. For the same reason, when making the compost pile, it should have "ripe" soil mixed into it—soil that contains bacteria and humus. Figuratively, this has the effect of a "leaven" or "sourdough" in terms of bread making. Having developed ripe compost, it is advisable, when taking it away, to leave a thin layer of the old pile on the ground on which to build up a new pile. Everything that will decompose into humus can be used for compost. One can use all sorts of plant refuse, straw, chaff from threshing, ditch mud, bracken, seaweed, potato vines, hedge trimmings, wood ashes, slaughterhouse refuse, horn, hoof, bone meal, and kitchen garbage; all inorganic substances such as broken glass, iron, and so on should be carefully removed.

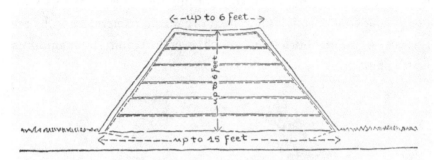

Construction of a compost heap

A model compost yard

Setting up a compost heap is done as follows: First dig a pit for the pile, 5 to 10 inches deep. If the pit is pure sand, it is best to spread a thin layer of clay over the surface; in a pinch, straw may be used. This should be covered, when possible, with a thin layer of manure or compost already decomposed; or if it is an old site, the bottom layer of a previous pile will serve the same purpose. What has been said about "drainage" and moisture of the manure pile is equally valid for compost. The structure and consistency of the compost should also be moist, but not wet.

Alternate layers of compost material and earth are then laid on the pile, and between these layers a thin sprinkling of unslaked lime should always be used. When the pile has reached a height of five to six feet, it should be completely covered with earth. The size of the pile should be kept to the following proportions: length, as convenient; breadth at the base, 13 to 15 feet; breadth at the top, 6 feet; height, 5 to 6 feet. Smaller piles may be proportioned correspondingly. The dimensions given should not be exceeded. If the material is plentiful it is better to start a new pile than to exceed the dimensions given.

Only in dry and hot climates are larger heaps preferable, since they conserve moisture better. In wet climates, narrower heaps are preferable, because they permit a better air circulation. This applies to manure piles, as well. The thickness of the earth—both the inner layers and the covering—depends on the nature of the soil in question. We should be guided by the fact that the fermentation of the pile is a life process, hence the pile must be allowed to breathe. It should have a skin that holds it together but does not isolate it. If the soil used is heavy and sticky clay, the thickness should not exceed 2 to 3 inches; when the soil is light, the thickness can be from 4 to 8 inches. It is important to note that soil from orchards and vineyards that have been sprayed with arsenic, lead, and copper preparations is impregnated with those metals, which are hostile to bacteria. Such soil is entirely unsuited for preparing compost. I have seen piles made with such soil that had not rotted after standing for 2 years.

It is of great importance that all compost material be moist. If, as with leaves and so on, this is not the case, the material must be moistened with water, or liquid manure must be poured on it when setting it up. Maintaining the proper moisture is one of the most important requirements of manure or compost heaps. In time, the compost maker acquires the necessary experience.

Insofar as possible, leaf compost should be set up immediately after the leaves have fallen and not the following spring, after they have been "washed out." While the pile is incomplete, the material spread out on it, for example the daily kitchen garbage, should be immediately covered with straw mats, sacking, reeds, fir or evergreen branches, straw, and such. In tropical regions, banana leaves are particularly suitable for this purpose. If the finished pile cannot be made in the partial shade of trees and shrubs, it needs a covering that affords a similar protection. When the pile has grown to about a yard in height, the treatment with the biodynamic preparations is begun. The preparations 502 to 506 are inserted at distances of about a yard apart, and the heap is sprinkled with preparation 507. What was said earlier about the treatment of manure applies here, too.*

This treatment speeds fermentation toward the formation of humus. In about 3 to 5 months after the pile was started, it is turned, and, if necessary, preparations are inserted again. After turning, the material is naturally mixed together and not layered.

The art of compost-making was better known in earlier centuries than it is now, even to many "experienced" agriculturists. In Flanders, there was a guild that had the sole right of collecting organic refuse during daytime and mixed with earth. The ripe compost found a ready market. Anyone outside this guild wishing to make compost had to build heaps in secret at night. In many areas of South Wales, people are still familiar with the old practice of building piles with layers of manure, rubbish, and quicklime and covering them with earth. For such piles one needs only to insert preparations given by Rudolf Steiner to add the proper "finishing touch." The Indore process of Sir Albert Howard—based on experience in tropical and subtropical conditions—is in many ways reminiscent of the biodynamic treatment, but not the use of the preparations.

---

* More detailed information on these topics is available in M. Klett, *Principles of Biodynamic Spray and Compost Preparations.*

All sorts of weeds can also be used in compost piles. One must take care, however, that this material goes into the innermost part of the pile, where, because of a lack of air, the conditions of fermentation are such that they destroy all the seeds. When turning the pile, the outside of the original heap should become the inner part of the new, and the former inside part now becomes the outside. Thus, weed seeds are destroyed in all areas of the pile. It is even better to make separate weed compost piles and let them lie longer, even up to 5 months, before turning, and a year and a half before use. The normal time of rotting of well-handled compost in the damp cool climate of Central Europe is from 8 to 12 months for all sorts of compost material. Cabbage stalks, which require about a year and a half, are an exception. Hence it is preferable to mix such material as this with weed compost. In tropical and subtropical climates, the fermentation occurs in about 3 months.

The final result is a fine, aromatic earthy compost, smelling like woodland soil (humus). Those who have more experience with the preparation of compost will quickly discover two things: 1) Making compost is an "art." 2) On every farm or garden where all the refuse material is carefully gathered, the amount of compost material is far greater than generally thought. But this is not true of gardening alone; in extensive farming, too, there are great reserves of compost when we include chaff from threshing, potato vines, ditch and pond mud, waste straw, turnip leaves not used for fodder, and other materials. On a mixed farm of 250 acres, with 14 dairy cows, 4 horses, a proportionate number of young cattle, and an intensive culture of grain, it has been ascertained empirically that the quantity of compost gathered in a year is equal to the quantity of stable manure gathered in 6 months, the latter being well mixed with straw. If we include the conservation of fertilizing values resulting from the use of the biodynamic treatment of organic manures and composts, we can confidently assert that, by composting the fertilizing materials not used before, a farmer will have, after the change of method, more than double the previous amount of fertilizer in terms of value and content.

It should be pointed out to agriculturists and fertilizer experts—who might think that biodynamic methods do not adequately cover the nutrient requirements of the land—that the full value of the biodynamic farmer's manure is conserved instead of being wasted, and the farmer thus creates a reserve. Biodynamic farmers can cultivate more intensively, because their organic fertilizer—the basis of all agriculture—is treated more intensively.

CHAPTER 6

Maintaining Living Soil
Cultivation and Organic Fertilization

I t is useless to focus attention on conserving manure and compost values if later on, in one's care of the soil, all the valuable substances are simply squandered through wrong handling. A series of steps should be carefully observed in this connection. Carting manure out to the field and plowing it into the soil, and so on are steps that should follow one another in rapid succession. When the manure is distributed in small piles or spread out on the field and left for any length of time, the important ingredients are in danger of washing away, drying out, leaching, and oxidization. When the soluble fertilizing substances are washed away in this way, and the nitrogen is completely gone, farmers wonder why they don't get the same harvest as they do with chemical fertilizers, or they are amazed that their manure was "poor," or they reach the usually satisfying conclusion that organic fertilizer is insufficient to meet the nourishment needs of the plant. But if they have done their best to dissipate the nutritive materials of the soil, they don't need to be astonished at poor and insufficient results. In any case, blame should not be placed on the organic fertilizer.

How, then, does the equation of nutritive materials stand? It might be assumed that one head of cattle weighing 1,000 pounds will yield around 13 tons of fresh manure, or that a cow of 800 pounds live weight will give 10 to 15 tons of manure if there is stable feeding all year long and a generous amount of bedding straw mixed with the manure. In the case of a mixed or diversified farm (70–75 acres of cultivated fields, 25–30 acres of pasture and hay fields), with stable feeding for half the year in the temperate zones, we can expect, in normal conditions, about 6 tons of manure from each grown animal.*

A ton of manure (without the conserving effect of the biodynamic treatment) includes the following important nutritive materials, shown in kilograms (1 kilogram = 2.2 pounds):

	Water	Organic substances	Total nitrogenous	Potash
Horse manure	713	254	5.8	5.3
Cow manure	775	200	3.4	5.0
Sheep manure	680	300	8.5	6.7
Pig manure	724	250	4.5	6.0
Mixed stable manure, fresh	750	210	5.4	6.0
Mixed stable manure, rotted	770	170	5.4	7.0
Liquid manure	982	8	2.8	5.5
Chicken manure	560	355	15.4	8.5

	Lime	Phos. acid	Silicic acid	Magnesia
Horse manure	2.1	2.8	17.7	1.4
Cow manure	4.5	2.6	8.5	1.0
Sheep manure	3.3	2.3	14.7	1.8
Pig manure	0.8	1.9	10.8	1.2
Mixed stable manure, fresh	5.0	2.0	10.8	1.2
Mixed stable manure, rotted	7.0	2.5	16.8	1.8
Liquid manure	0.3	0.1	0.2	0.4
Chicken manure	24.0	15.4	35,2	7.4

* To estimate weight of piles: 1 cubic yard of fresh strawy manure weighs about 1,100 pounds; 1 cubic yard of well-rotted, not-too-wet manure weighs 1,300 to 1,400 pounds.

The total of nutritive materials taken out of the earth in a 3-year crop succession—4 tons per acre of potatoes, 1¼ tons of winter wheat, 1¼ tons per acre of summer grain—will show the following tabulation in pounds-per-acre on the assumption that straw and vines have gone back into the fertilizer:

	Potash	Phos. acid	Nitrogen
Potatoes	134	15	33
Winter wheat	17	27	67
Oats	14	25	61
Totals	165	67	161

Based on this table of decomposed stable manure ingredients, it is shown to be necessary (where fertilizing occurs once in 3 years) to use approximately 12 tons of manure per acre. This is true provided we base our calculations only on the bare figures of quantities extracted and substituted, and that we ignore the question of soil quality and its reserves of the necessary ingredients. Such a manuring per acre is, in fact, usually given on mixed farms. This total should suffice if we consider the fact that certain losses are made up by the previously described, careful treatment of manure. In terms of cows, with 6 months stabling, this means, 2 head of mature cattle per acre of cultivated field, and is the proportion usually attained on small farms.

Experience has shown that, on large grain-growing farms, the proportion is generally not up to this standard. The situation can be improved by crop rotations that are not exhausting to the land, and by using catch crops. Planting legumes is especially valuable for meeting the nitrogen requirements, yet without counting the "reserves" in the soil.

This rough calculation sheds light on an important problem of whether certain farms under consideration are in a biological state of health. For instance, which of them are healthy as a result of the equilibrium of their cultivated and grass areas and crop rotations? And which ones, viewed from a biodynamic perspective, are going in an unhealthy direction owing to one-sided practices, and are thus farms to which "correctives" cannot bring a lasting cure? The author has actually seen farms that, although having only 1 head of cattle per 7 to 10 acres, were able to keep going. But in such cases, all the finesse and skill of experienced farmers are required to preserve the farm's "organic foundation."

We can clearly deduce from this the amounts of manure needed to adequately fertilize a piece of land. If care is taken that the full fertilizer values are retained in the manure, no one who follows this recipe needs fear robbing the soil or risking poor harvests. It is the primary goal of the biodynamic agricultural method to educate farmers to this high standard of manure utilization.

Those who read this exposition of the biodynamic method may think that only a farmer with the necessary amount of cattle is in a position to start using the biodynamic method of agriculture. Our reply to this would be that, in fact, it is true that no farm can have real value unless it is based on a proper proportion of pasture, hay fields, and cattle to tilled fields. The biodynamic method of agriculture was *not* designed for a farm without cattle and for which no organic fertilizer can be purchased. A farm without cattle represents biological one-sidedness contrary to nature. Likewise, a mere cattle range such as often exists in hilly regions or in places where the soil is too heavy also represents something essentially one-sided and needs a regulating, balancing factor. Those who object to the state of balance advocated in biodynamic agriculture have a complete misconception of what a farm should be. Making things grow on a piece of soil is not necessarily farming—it might merely be destroying the earth's fertility.

Returning to caring for soil and manure, biodynamic manure— quickly decomposed and usually easily spread with a shovel—should be spread over the ground the moment it is carried to the field, and then immediately plowed under. If it is desirable to cart the manure to the field at a time when other work is slack—on a rainy day, for example—the entire pile may be taken to the field where it is to be used and piled up and covered there. This may be considered a process similar to turning a manure pile. Once the manure is spread, however, it should be plowed under within 3 hours to retain its full value. Neglect of this point can lead to great disappointment and loss of crops. Indeed, one needs only to walk over farm fields nowadays to see the sins committed against nature and healthy agriculture.

A slight exception to the rule of plowing in manure immediately is when manure is carted onto frozen ground and freezes, which saves it from disintegration for the length of the freeze. Spreading manure without immediate plowing it under is also allowed under certain circumstances on pasture and hay fields; it might be necessary when working with very heavy, wet soils. In the case of pastures, a good harrowing should have been done previously so that the field is able to quickly take up the manure solution as it thaws.

When the manure is plowed under, care should be taken not to bury it too deep. It should continue the process of transforming itself into humus in the ground; since this is a biological process, to do so requires air, life in the soil, and moisture. The most, intensive life activity in the soil exists in a surface layer of about 2 to 6 inches deep. The layers below this are relatively lifeless and should not be disturbed, because there is less aeration in them and thus considerably less possibility of transforming organic substances. It is therefore very important not to plow organic fertilizer into the soil too deeply, otherwise it will fail to develop its full effect.

Deep plowing might in fact loosen the ground, mix it, and bring up hitherto unused layers of soil; yet, insofar as the biological state of the soil is concerned, such plowing offers no advantage—in fact, quite the contrary. By means of deep plowing, the living upper layer is buried, covered up, and isolated, and life processes are brought to a standstill. Manure, for example, that is plowed under to this lower level, remains unaltered for a long time in heavy, cold, and wet soils. Months later deep harrowing may bring to the surface the unaltered fragments. Furthermore, the lower, inactive layer that is brought up by deep plowing, will usually need several years to become permeated with soil life, bacteria, etc. In the meantime, it is not in a biologically active state. It continues for a long time to lack the proper crumbly structure, and the surface tends to be incrusted. Moreover, the pressure exerted by the new, more compact upper layer over the crumbly layer, which has been plowed under, might lead to the formation of two separate levels, causing water to stagnate and disturbing circulation. Dammed-up moisture is harmful to germinating seeds. The need to plow more deeply for root crops and potatoes than for grain is obvious. We are concerned here with fundamental principles of obtaining the maximum value of the manure, which in practice must be adjusted to particular crops and types of soil.

Let's assume that, up to this stage, everything has gone well— careful piling of organic fertilizer, covering, inserting the preparations, moving it out to the field, quickly plowing it under to the correct depth. There are, however, still a number of factors remaining that will determine the degree to which organic fertilizer has been properly used. Air and moisture are not the only requirements; we just pointed out deterring influence of stagnant moisture. Such a condition can be remedied partially through proper drainage, an absolute prerequisite for healthy biological conditions. Most important, however, is the maintenance of capillary action in the soil. Any measure that disturbs this becomes a hindrance to getting the most from manure. The most generous manuring, if accompanied by incorrect soil cultivation, results in only half the value.

While investigating complaints of insufficient yields, I have often had to point out mistakes in cultivation rather than poor or incorrect manuring. Biodynamic farming cannot be done without correct cultivation. In cases where this is neglected, farmers should not blame the biodynamic method but rather their own errors.

The causes of mistakes in cultivation are often difficult to bring to light. Although often neglected, the most important and well-known points to keep in mind are care in the plowing and harrowing of wet soil and to avoid bringing up clods or smearing and closing off the surface. Care must be taken to use the roller at the correct time—when the ground is becoming dry—to retain capillary activity and prevent surface incrustation. Soil that is alive and kept permeated with organic processes inclines of its own toward a crumbly structure. Those who maintain this sort of a farm or garden through correct cultivation at the proper time have become friends of the biodynamic method of agriculture in regard to soil improvement. Often it means having the patience and courage to wait a few days more before working soil that is still too wet and heavy. The biodynamic point of view teaches this caution.

A complete and quick change of all plans might be required to roll the soil at the necessary time so that it doesn't dry out and the surface doesn't close with a crust. With care, experienced farmers can always prevent the incrustation of the soil surface. The much-maligned "slow" farmer knows that starting the "right work at the right time" brings the harvest. Plowing and manuring are only a preparation, but taking advantage of the correct state of the soil and weather is what generally determines the harvest. If young budding agriculturists were only given more practical drilling on this point, many merely passable lands might be turned into first-class farms. But it can be frequently seen how countless valuable factors in the soil are lost through lack of knowledge and care. This is true, above all, in "young," newly settled lands, where no old, soil-wise husbandry exists. Half of the dreaded condition of erosion in subtropical and tropical regions is caused by lack of knowledge and care.

There is a distinction between crop rotations that conserve humus and consume humus, and between humus-conserving and humus-consuming weather conditions of different years.* Crop

---

* Wet and cool years are more humus conserving than hot and dry years.

rotation overbalanced on the side of grains, for instance, consumes nitrogen from the soil, whereas legumes conserve it and, of course, return nitrogen to the soil. They also act on the weathering of the soil, and thus free new reserves. A study of the comparative action of various cultivated plants in opening up the soil is very enlightening. In dealing with "biological weathering by living organisms," E. Blank gives important figures.* For example, he shows that various plants take up entirely different amounts of mineral substances from one and the same soil.

The following show the amounts (in grams) of mineral substances in specific rock materials brought into solution and taken up by various plants:

Number of plants	From variegated sandstone	From basalt
3 lupine plants	0.60	0.75
3 pea plants	0.48	0.71
20 asparagus plants	0.26	0.36
8 wheat plants	0.027	0.19
8 rye plants	0.013	0.13

Legumes have thus approximately a sevenfold to sixtyfold stronger effect in breaking down the soil than the grains. This fact, along with proper crop rotation and cultivation, as well as the fixation of nitrogen, points to the significant benefits of the *Leguminosae,* the use of which at certain times should not be omitted from any crop rotation. Only by using legumes can a significant proportion of the natural reserves be brought into use for fertilizing the ground. This, together with the handling of manure and compost, constitutes one of the most important aspects of the biodynamic agriculture methods. Too frequently, however, planting legumes causes an excess of nitrogen, which in turn hinders the fruiting of plants and causes them to tend toward leafiness. Thus, legumes—i.e., peas, beans, clover, lupine, alfalfa, serradella,

---

* E. Blank, *Handbook of Geonomy,* vol. 2, p. 260.

and so on—must always form part of a farm's healthy crop rotation as a therapeutic factor; but balance and moderation must be maintained, too, in the case of legumes.

CHAPTER 7

How a Conventional Farm
Becomes a Biodynamic Farm

It's difficult to discuss the general principles of converting a farm to the biodynamic system without going into considerable details. As to the important question of planting, we see that every region, soil type, climate, and particular market condition requires its own type of crop rotation. In every case, however, a legume should be grown among the crops every four years. This is often done by sowing clover with the summer grains or planting beans and peas instead of summer grains. Crop rotation extending over a cycle of 5 to 7 years (often the rule in northern countries) has a more beneficial effect than the more "intensive" 3-year rotation.

The following is an example of a somewhat overly *intensive* program for a biological farm: *year 1,* hoed crops (potatoes, beets, and turnips); *year 2,* winter grains; and *year 3,* summer grains. An even better rotation would be to sow clover among the third-year grains, which would then remain into the fourth year. In heavy, wet soil near the sea or in a damp, hilly climate, the hoed crop during a moist autumn might ripen too late to prepare the land for planting winter grains. In such a case, rotation should be as follows: *year 1,* hoed crops; *year 2,* summer grains; *year 3,* legumes; and *year 4,* winter grains. The effect of stable manure used on a hoed crop remains in the soil for the following summer grains. This is supplemented in the third year by planting legumes, which produces good winter grains with less danger of lodging from moisture. Other variations are also possible.

Another improvement comes from growing grains with inter-crops and green manuring or by planting legumes after grains. Legumes are especially important for light, sandy soil. Just as in forestry, all monoculture has proved itself biologically unsatisfac-tory, just as growing unbalanced field crops has been proved unsat-isfactory. By using grain plus something such as oats with vetch or field beans, the beneficial effect of the simultaneously growing legumes restores biological balance. On sandy soil, an inter-crop of serradella is to be recommended. Green manuring with vetch can be profitably made use of in a heavy soil.

Many will say: *Of course! We know all this without biodynamic agricultural methods.* But might ask: *Does anyone actually act according to this knowledge?* The point is to actually follow these principles. We have made it fundamental to the biodynamic method of agriculture to energetically practice these principles of procedure to improve soil fertility. While familiar in themselves, however, they are generally neglected.

We should also point out another problem. In a climate of moist, rainy summers, growing grains mixed with other crops involves certain difficulties. The grain is ripe, while the inter-crop (vetch, perhaps) is still green. When the field is cut, the grain dries more quickly than the legumes, which are still moist, and this can have a negative effect on the drying process. Feeding a green grain crop where there is a heavy soil (in Holland, for example) is too expen-sive as an alternative procedure. We need to give special consider-ation in each particular case. For example, the farther we go into Europe—to the drier east or in other area of the world with dry and sandy areas—the more valuable it is to grow inter-crops. Even if that inter-crop is to be used only as fodder with grain straw, this sort of planting is advantageous, since it provides protein require-ments in home-grown feed.

With green manuring, we are at a disadvantage in moist climates near the Atlantic by having very little frost during winter. The simplest way to green manure is plowing the plant mass under after frost has worked into it. But plowing plants into the soil while still green *before* frost means that the unrotted plant mass will take too much oxidizing force out of the soil and thus diminish possible bacterial life. This danger is greater in heavy, firm soils than it is for loose, airy soils.

Following biodynamic principles, it is also better to add green manure to the compost pile to produce a neutral humus. To do this we need to gather the material cut green, or even after it has been frozen, and compost it in layers according to the approved method. In this way, the soil will get humus that it can take up rapidly. This green fertilizer compost will, under favorable conditions, be ready for use by spring. The roots are plowed under. In moist climates, however, we have wet soil that makes autumn cutting and gathering green material difficult. In this case, therefore, the entire plant mass must be evenly turned under by shallow plowing, followed later by the deeper winter furrow. If there is a good, crumbly soil structure, and if preparation 500 is applied at the time of plowing under, this procedure might prove satisfactory. On muddy firm soil, however, this is not advisable.

Planting legumes following grains is useful in intensive farming. Beans, vetch, lupine, and so on follow a grain harvest. This brings nitrogen into the soil, and the ground remains covered for the rest of the summer and is protected against loss of moisture. The resulting harvest is a high protein fodder relished by the animals. The important thing is that immediately after the grain harvest the stubble be peeled off and the ground harrowed, sown, and rolled. Only in this way can the beneficial, crumbly structure of the soil be maintained. Every day and even every hour that is lost putting in the crop that is to follow—after clearing the field—means the loss of water and soil structure. Experienced farmers know that, *within three hours after mowing the grain, vetch must have been sown already.* The rows where the grain sheaves are set up should be planted separately, later on. Such a procedure gives an idea of what quick cultivation means.

Questions of crop rotation and the possibility of using inter-crops or after-crops constitute important problems when converting a farm to the biodynamic method of agriculture. What's the best way to carry out this conversion? First, we should not approach farming in a random way but make a clear plan. It has already been said that it takes several years to rejuvenate soil whose organic structure has been disturbed, and that the biodynamic effect can be fully developed only after this period of time—in other words, while in the process of crop rotation, manure is applied the second time on the same parcel of land. Thus, any plan for converting a farm will require at least two complete crop rotations.

The first step in this plan is to establish a manuring schedule. To the extent that the organic fertilizer is available in sufficient quantity, the conversion can be carried out. *The first practical measure, then, is the careful treatment of the manure.* A farmer who has recognized the value of this work will be ready for further suggestions.

The second practical measure is to carefully gather and handle all the organic masses that can be turned into compost. Treatment of the fields with preparations 500 and 501 begins during the first year of converting the parcels where legumes are growing and wherever manure or compost can be used immediately. Once the prepared biological fertilizer is applied, the first stage of the conversion is complete.

Every farm that uses both regular crop rotations and a set number of cattle can be converted to the biodynamic method in this way. Farms without cattle cannot be developed with the biodynamic system. In gardening, where there can be no cattle, buying stable manure or other organic matter is the only way.

When organic fertilizers are available, but in insufficient quantities, they can be supplemented by buying horn and hoof meal (for nitrogen) bone meal (for phosphoric acid), and so on. This does not cost more than mineral fertilizer, and if the material is composted first, it improves the soil.

Soil conditions, local demand, and available markets determine crop rotation. No one should be advised to grow peas if they cannot sell them. But farmers can be advised to grow a fodder legume that they can use for feed.

With the second stage of the conversion, farmers should increase the number of cattle and, thus, manure production. They should improve the fodder and gradually change, insofar as possible, from the intensive feeding of concentrates to home-grown feeds, which are healthier for cattle, the manure from which improves the humus of cultivated fields. The increased milk yield obtained by feeding concentrates is generally not economically sound, given the other side of the ledger and the expense of replacing cattle lost to sickness and aborted offspring. Milk production is needed commensurate with the healthy condition of one's particular type of cows. When calculating the economic value of a cow, both milk and calf production must be considered. The average cow with many calves is more economical than a cow that produces a lot of milk but few calves. She is also more valuable because the breed of her calves is improved. In other words, a cow that is not exhausted by over-milking transmits more life energy and organic reserve strength to her offspring.

The issue of sterility belongs presently to the farmer's daily problems. Much is said about Bang's bacillus, the dangers of infection through the bull, especially from a village-owned bull or one owned by a farmers' association. There is great danger when illnesses are introduced to a "clean" stable by newly purchased animals. However, we know from the biodynamic perspective—and even more so from practical experience gained over a decade—that an infection requires two organisms: the organism being infected and the infecting bacillus. From recent reports in the magazine *Demeter,* we know that it is possible to combat disease in the stable through the sanitary handling of the livestock and sanitary feeding and careful treatment of manure.

Thus, it is interesting that—even in a traditional cattle-raising country such as Holland—it was "discovered" that the issue of sterility might not be simply a matter of infection; something else might also play a role. The following paragraphs are a translation of parts of articles in the popular Dutch farming journal *De Boerderij* (May 12 and July 28, 1937).

> The causes of sterility vary. These causes are also different from those in the public eye 10 to 25 years ago. This change of causes can be attributed to changed farm conditions.
>
> During the period behind us, people thought almost exclusively in terms of infection. An ailment was thought to be connected mainly with epidemic abortion. The bull transmitted the disease bacilli or even became sick himself. Thus, the cows failed to calve. Improvement and early recovery depended on disinfectant treatments...but the result of such treatments was not always positive; much still remains unexplained. Moreover, the condition of sterility tended to spread, just as today in the case in epidemic abortion.
>
> This development has, in fact—in a far more comprehensive sense—contributed much to the consideration of the sterility problem from different perspectives.
>
> At present, we are already able to ascertain, without exaggeration, that infection plays only a minor, even negligible, role in these problems. This does not mean, however, that those who raise cattle should neglect the dangers of infection, but that they should not attribute inordinate importance to it. By doing so, they run the risk of overlooking the main cause of the ailment. They might rinse, scrub, cleanse, and apply ointments as much as they please, but in spite of this their establishment suffers and the cows remain sterile.

The report continues to speak of pedigreed bulls, considered "first-class" animals:

These animals are primed for so-called exposition standards. This can damage the animal fattened in this way and result in partial impotence. Of course, the bull might recover through diet and exercise, but the best months are lost.... Sometimes this fertility decrease is caused by a lack of vitamins; this might be the case with both bulls and cows. A bull in a dark corner of the stable, with food too rich in albumin and insufficient activity, has little chance of breeding vigorous progeny.... This applies also to cows. Here, too, constant stalling and the type of fodder can influence the animal's reproductive capacity. Pasturing certainly brings a change, but such improvement often comes too late and might cause one's enterprise great losses. Measures are needed to eliminate at the right time the hindering influence of too much stalling, overfeeding, and artificial stimulation of the milk yield. Care is needed to preserve the balance between productive and reproductive capacity.

The report goes on to state that experiments with artificial fructification have resumed:

The fact that this kind of technical method is applied to cattle again today indicates the far-reaching, unnatural situation of current cattle raising.... This is the primary cause of the ailment. We will have to return to natural cattle raising. This must be the primary goal of our endeavors—also in breeding.... *When they are bred only for milk production, why should the animals become so weakened that they are unable to give birth regularly to even a single healthy calf?* The sooner this changes the better! [emphasis added]

This description comes from a country famous for its cattle raising!

Here we must keep in mind that natural life conditions always offer the best guarantee for a strong constitution, and that the consequence of a strong constitution is stronger resistance capacity. The more we tried to gain through high milk productivity, early maturity, and high fat content, the more we had to consider factors resulting from sterility. A more recent view of the problem diverts it in another direction. Hence, we are no longer justified in seeking the cause of sterility in contagious diseases of cows and horses by continuing to slavishly and increasingly follow outmoded concepts.

During a certain period, hereditary influences were given too much attention. Deterioration was attributed to heredity, whereas wrong feeding was actually the cause.

Sometimes pasture grass did not contain the right ingredients, most likely because of excessive manuring with nitrogen. High albumin content acts detrimentally. It is known that rich grass easily causes a form of meningitis, a disease increasing greatly in Holland.

By calculating the market value of manure, a surprising discovery is made; the normal combination of pasture and arable land is economical despite lower wheat production for the market—and this in a region not devoted primarily to dairy farming. In any case, cattle must be brought to a *high* (though not *forced*) standard by means of careful breeding and selection. Huge cows prepared especially for exposition through a daily feeding of 24 quarts of milk, oat flakes, and zwieback are not more capable of healthy breeding than are large, long-legged cows with all sorts of hidden defects. If such animals have been purchased and brought into normal conditions, they refuse their normal diet, become emaciated, and need considerable nursing to bring them back to normal. It is a misfortune for good breeding regions that the best animals are sold for export, leaving only second-rate animals at home.

The most practical ratio between meadow and arable land, and the calculation of the most advantageous acreage of each for the quantity of manure needed, should be carefully worked out for each individual farm. To attain the best average yield requires several years. It also depends on the capital reserves of individual farmers, which should be called upon as little as possible. When ample financial reserves are available, the same goal can be reached in three or four years, which might take other farmers 8 to 10 years of gradual building. If carefully done, the conversion will not fail. The high premium for knowledge that makes this assurance possible has already been paid by the pioneers of the biodynamic movement; today their experience is available to all.

During the period needed to achieve the proper proportion of pasture and arable land, the proper proportional increase of quality in cattle can also be achieved by intelligent breeding. Related to this, an important fact must be stated; high-bred cattle, having already reached the apex of improvement, cannot be further improved. *If they change at all, it is for the worse—they degenerate.*

When it comes to breeding, strong, average cattle, capable of further development are the most desirable and best. We have observed that local types, developed under *traditional* methods, offer the best possibilities for further improvement. The importation of breeding cows should be limited to the absolute minimum necessary. Improvement of the breed can be attained by buying a bull from outside. Every *purchased cow,* brought in from outside, might bring diseases with it. A stable from which contagious abortion has been eradicated might nonetheless suffer from reinfection brought in by a purchased cow. This author had the best breeding results when introducing young bulls, not quite a year old, raised on soils that were a little poorer than his own and from a climate that was a little more severe. If the points and characteristics of the animal were good, then further improvement was possible by improving its environment. Observant farmers quickly discover that their own breeding, when properly done, always gives the best results.

Let's summarize the steps of conversion:

1. Consideration of what is *desired* and what can be done
2. Proper care of manure and strict application of preparations
3. Setting up healthy crop rotation and improving the methods
4. Improving the quality of manure

5. Improving feeding with home-grown feed
6. Improving the herd as a whole

All further procedures with the biodynamic method of agriculture depend on the judgment of individual farmers. Steps 1 to 5 are absolutely necessary to attain the best results. Maximum results are attained by the intensive work of farmers and, under certain circumstances, by using additional capital. The questions of seeds is very important. We already remarked that biodynamically raised seed has proved its resistance to plant diseases. Hence, biodynamic farmers must gradually raise such seed themselves on their own farms. For example, it is always practical for a moderately sized gardening establishment to grow some its own vegetable seeds, For the general purpose of this book, it is unnecessary to give more details about seed growing, although it must be mentioned that seeds for hothouse cultivation should always be produced on open fields.

Seed culture without strict selection is impossible, even when so-called mechanical cleaning and sorting are practiced. If, for example, potato seed is being selected, the stand of plants must be observed throughout their entire period of growth. The strongest and finest blossoming plants should be marked. Of these only those surrounded by healthy plants should be used. Finally, those that eventually are best are dug up before harvesting the others. This makes extra work but repays the effort.

Another issue has to do with putting up protective hedges as windbreaks. This is important for pastures, especially those used for young cattle. We have observed further that the digestive activity of cattle is beneficially stimulated when they eat leaves from shrubs and trees. For example, a little "nibbling" on the leaves of the hazelnut increases the butterfat content of milk.

There is one question that arises repeatedly in connection with converting a farm—namely, that all these things require more labor, which greatly adds to labor costs. Thus, it is said that conversion is expensive and does not pay economically. In fact, considered superficially, extra labor is needed—for example, to gather compost, turn manure or compost piles, and, in the long run, make changes to cultivating methods and crop rotation. Growing mixed crops means extra work, and so on, but these expenses should be compared to the amount of time we consume in purchasing; hauling and spreading mineral fertilizers; spraying copper sulphate or similar materials; working dead, encrusted soil; purchasing and hauling concentrated feeds; and even constantly trading in cows. This latter might actually be a pleasant occupation, generally carried out over a cup of coffee (or glass of beer), but it reveals the fact that something is wrong in the stable. When we count all the time spent in this way, the whole matter assumes a different complexion.

In this connection, the author is able to cite testimony of those actually engaged in farming. One farmer reports that, on one acre, he previously expended 19 horse hours of work and 27 human hours of labor each year. The conversion required 21 horse hours and 30 human hours for the same amount of ground because of the increased work with manure and compost handling, hauling, setting up, covering, and turning. The horse hours seem high, because this farmer had to travel more than a mile to reach his fields. To balance this, the farmer puts the higher fertilizing value of his manure on the other side of the ledger. When ripe manure is spread, we discover the first reduction of labor required. It is so well decomposed that it can be spread out directly and quickly with a shovel. Manure is loaded from the pile directly onto the wagon, spread-broadcast from it, and then plowed in. Previously, at this point there was another step—the manure was taken from the big pile and set up in little piles on the field, which were then spread out. This involved additional losses of nutritive materials, especially when several days pass before the manure is plowed under. It is advantageous for the farmer to plow the manure under immediately after spreading.

Controlled experiments in Holland have shown that, if the yield of immediately plowed manure is figured at 100%, then the yield from manure that has lain on the field for three days is only 86%, and the yield from manure that has lain on the ground for several weeks is correspondingly less.*

We have already spoken of the fact that the biodynamic procedure helps to bring lightness to the soil and develops a crumbly structure. The farmer whose experience we cited has also had this result. It became clearly evident in field work, harrowing, hoeing, and so on. Hand hoeing potatoes on this farmer's place formerly required nine hours per acre; now it takes seven. He further reports that, in the crumbly soil, weed seeds can be eradicated more quickly, generally with just a cultivator. The time needed to stir and spray preparations 500 and 501 is balanced by the previous need to haul and spread mineral fertilizer during the busiest period of the year.

Another farmer operating a large farm reports that it previously required 4 human days of labor per acre for hoeing sugar beet fields, but now, with looser soil, the same work takes 2.9 days. He confirms the figures for potato hoeing as 7 hours now in contrast to the former 9 hours. On a farm of 437 acres, he has been able to do with one less team. He reduced not only hours, but also the fodder required by his horses.

On larger farms, it is essential to make one or more people individually responsible for the biodynamic measures. Only in this way can we be sure everything will be done correctly at the proper time. Such workers should be specially training on biodynamic model farms. We should mention here a social consequence of this new method; it requires a greater interest in natural phenomena by farmers, and thus raises them to a higher concept of their calling. Moreover, because the work is intensified, it offers the possibility of resettling more people on the land.

---

* Note: In North America, widespread use of the mechanical "manure spreader" (while eliminating the aforementioned intervening stage of the little piles) gives a nearly corresponding loss of fertilizer values, because the plants are unable to fully utilize the raw, unrotted manure that is typically spread by this method for plowing under.

This author has had an opportunity to observe several homesteading experiments, most of which have fallen apart because those involved were not trained well enough in their calling, nor did they have enough interest. A certain amount of exact knowledge is essential. A lack of interest shown by those who have been industrialized cannot be remedied merely by settling them on the land, teaching them the laws of nutrition, and promising high crop yields.

The right foundation is created only when homesteaders have an inner relationship to their work—when they learn to survey and comprehend the wholeness of the life processes of an agricultural organism. They will then also love it as one can love only something that is alive. With refined and wakened senses, they will see every single reaction of the soil, plants, and animals in the large interconnection that in itself indicates health and growth and represents their own future.

Enthusiasm and *good will* on their own guarantee no more success than does the presence of mere bodily strength and energy. We have observed this in most cases of rural resettlement. In one case there was abundant enthusiasm but no working power; in the other, work power with no enthusiasm or will to work. A homesteading experiment on a grand historic scale that was only partially successful was the settling of the North American West during the 19th century. Conditions in those areas today offer a good example of what a general disregard for the laws of "biological totality" on farms can produce. By contrast, these healthy conditions remain today in sections of America where a good farmer population, such as the Pennsylvania Dutch, settled.

In addition to the plan for converting the fields, one must also establish a fodder plan. Because our goal is to manage with home-grown fodder insofar as possible, its planting must advance in the proper proportions. This is done on the basis of knowledge of soil-conserving crop rotation—above all, correctly timed planting of clover, alfalfa, and mixtures of legumes with summer grains.

While we want to suggest several possible programs, we should remind the reader that this theme cannot be treated exhaustively here. The ideal feeding is pasture during the summer with clover and grass, and hay during winter. Small farms—above all, those in hilly regions—often come closest to this, with clover and grass pasture, beginning in the early summer, turnips in the fall, hay and straw in the winter.

One sugar beet farm feeds a considerable amount of clover, straw, and legume hays (mechanically dried). Another plan: **May to September**: pasture and straw; **September to October**: green corn (maize); fodder cabbage (planted after grain harvests in a sandy soil), some potatoes, straw; **December to May**: hay, mixed straw, fodder beets and the like, and legume hay.

The program of a grain farm: **May to September**: red clover, alfalfa, straw (when cattle are stabled); **October to November**: turnip leaves, straw, hay, clover hay; **Winter**: marigolds, chopped turnips, a little distiller's mash, hay, straw, bran. Such examples could be multiplied extensively and vary according to soil and climate.

Another important point in the conversion of a farm has to do with the value of plowed-up meadows and pastures. If the soil can be said to have rested after only a year of legumes, how much more can this be said of it when it has lain as pasture or meadow for a longer period? In such a case, the result is a biologically quickened soil that, for the first one or two years after plowing-up, needs no fertilizer and provides very good yields.

When good yields do not result, we find the cause in a common mistake. A meadow has been plowed up, and the grain or potatoes planted in it have produced a poor crop. Why? The answer is that the pasture itself had become poor and thin *before* it was plowed up. It had perhaps been poorly cared for without compost or was wrongly pastured or cut.

A meadow also needs care; it needs compost now and again. We have found that compost has a markedly better effect than manure on pasture land. If no compost is available, then at least the manure ought to be given in a composted form. The more earthy it is the more easily is it taken up by the grassy plots. There is considerable loss if the manure—especially manure full of straw—remains lying on the grass too long and is dried, leached, or washed out. *The surface of a pasture or meadow, as well as a plowed field, must be able to "breathe."* It needs aeration just as much as does the soil around fruit trees or, for that matter, all soil. A matted-down and closed-up surface means a retrogression of the good grasses. The sour ones stay, and we can look for an increasing acid reaction of the soil, the disappearance of clover, and, in moist climates, increased mossiness of the ground.

Hence, pastures and permanent hay fields must be harrowed occasionally in the fall or early spring. If the soil tends to "mat," it must be harrowed once, deeply enough to make the soil visible. Where the grass is growing well and there is enough moisture, it recovers quickly. Naturally, the grass must not be torn out by the roots; special pasture harrows are used for this purpose. When earthy compost or manure is spread on the freshly harrowed grass, or even harrowed in, it is taken up immediately by the soil.

Correct grazing or mowing is a part of the necessary care of a meadow. A freshly sown meadow needs a definite amount of time to grow so that all the grasses take a good hold. A profitable pasture or meadow should always be sown with a mixture of various grasses and clovers. For particular conditions of soil and climate, a variety of six or seven grasses should be chosen, including high-growing and low-growing types, as well as four or five kinds of clover. If this mixture grows, during the first year the creeping grasses will grow more quickly than the long-stalked ones. After several years, the latter will predominate. With time, a balance is struck.

The essential point is that the important low-growing grasses and the clover are not eradicated at the beginning. This happens if the newly sown pasture is grazed early during the same year when the

Examples of Mixtures

For a heavy moist soil	Pounds per acre
Lolium perenne, English rye grass	9.0
Lolium italicum, Italian rye grass	10.6
Festuca pratense, meadow fescue	10.6
Poa pratense, meadow grass	3.5
Festuca rubra, sheep fescue	3.5
Phleum pratense, timothy	3.5
Trifolium repens, white clover	12.3
Trifolium hybridum, alsike clover	5.3
Trifolium pratense, red clover	3.5
Other possible mixtures	Pounds per acre
Festuca pratense, meadow fescue	5.3
Lolium perenne, English rye grass	7.0
Daktylis glomerata, cock's foot	3.5
Poa pratense, meadow grass	15.8
Cynosurus cristatus, colsted dogstail	3.5
Phleum pratense, timothy	1.7
Trifolium repens, white clover	9.0
Trifolium pratense, red clover	3.5
Medicago lupulina, yellow clover	3.5

ground is still wet. It should be mowed early the first summer, and grazed later. In warm or coastal regions, where it does not freeze in winter but is moist, and where the cattle remain outside during winter months, there is danger that the wet pasture might be tramped down too much, in which case the pasture should be divided into sections and the cattle moved from one to another, giving each part a pause for rest and recovery.

The important thing is always to keep a corresponding balance between high- and low-growing grasses. This varies from soil to soil and can best be learned by inquiry at the nearest agricultural experimental station or farm bureau.

TABLE OF FODDER VALUES

Plant	Protein %	Lbs. of fresh starch per 100 lbs. of green plant mass	Protein %	Lbs. of hay starch per 100 lbs. of hay
English rye	1.3	10.3	3.3	22.5
Italian rye grass	1.3	11.4	4.9	36.6
Timothy	1.0	14.0	3.2	29.1
Average of all Sweet grasses	1.5	13.7	4.0	30.2
White clover	1.9	8.8	4.9	32.1
Red clover	1.7	10.0	5.5	31.9
Alsike clover	1.7	7.9	5.6	29.8

The following figures are correct when the plants are in blossom at the time of cutting. Farm tests show further that, before the conversion, 100 pounds of starch are transformed into 169 pounds of milk with 5.21% butterfat, and after the conversion to the biodynamic system, 100 pounds of starch are transformed into 215 pounds of milk with 6.73% butterfat.

If we are not grazing but mowing, then the first cutting should not be too late, as often happens with the mistaken idea that we want "a lot." Grasses and clovers have the highest content of feed and nourishment values just before and when in blossom, which is when they must be cut. Later, they give more "straw" but less nourishment. If they are cut later, the possible second or third cutting is poorer and, the balance of the grass mixture changes to the detriment of the lower-growing grasses. Properly timed cutting, and later the grazing of hay fields, is very important, and fertilizing is equally important. Fresh manure and liquid manure are absolutely harmful and have the same negative effect. The clover gradually disappears, and the meadow becomes sour. Only well-decomposed, earthy compost helps (compost preparation is described in chapter 5). Only this will keep the clover growing. The disappearance of clover must be considered a danger signal. If weeds increase, and if grassy areas, instead of being green at blossoming time, are white or yellow with weeds—as we see rapidly increasing on Swiss pastures—it means the soil is sending out an SOS. Now it is time to plow up, manure, and aerate the soil. Just compost on the grass is no longer enough.

Farmers often plow up a meadow when no clover is growing, becomes sour, and produces only strawy grass. They plant grain and are disappointed when the "conversion" doesn't work. Thus, we see that meadows and pastures do not represent a reserve of soil fertility unless they are in good condition. Only then can we avoid fertilizing for a couple of years. If the clover has disappeared and the grass is poor, we must manure the plowed-up meadow immediately, since we have poor soil in need of improvement. It is important not to wait too long to plow. So long as we can keep wild white clover (*Trifoleum repens*) growing, all will go well. This clover has special significance, because it is drought-resistant and can improve the worst soils. For drought areas, lespedeza has also been shown to be valuable. Here, we must also cite the value of "short layers" (temporary pastures) in crop rotation.

Agricultural crop rotation, extending over eight-year periods, might prove practical. If so, a grass and clover mixture should be sown. This remains for four years as a hay field, cut several times each year and pastured in the fall; it should be given compost between the third and fourth years. Then it is plowed up and normal crop rotation of field crops begins. We consider such rotation the best and most soil-conserving, and with proper care it also gives a high yield of hay. This system is especially suited—even essential—for heavy soil in a moist climate to counteract souring and the spread of moss. This author has used it with good results on his own farm in Holland.

Plowing up of meadows is ideally done in autumn so that sods can break up well during the winter. Sowing summer grains follows naturally during the second year. Plowing up at the end of August is even better. This is followed by several successive harrowing, then comes winter plowing. In this way an especially mellow soil is obtained that, in late autumn plowing, would hardly have had time to develop. The extra crop of grass can be harvested in the high mountains in autumn, but has less value than cultivating the mellow soil mentioned. If the stand of grain was moderate, a hoed crop may be planted with manure. One might also plant peas and then beans immediately after plowing up. This is advisable when the preceding grass was in poor condition, and it is good to use some compost or ripe manure with the beans during the second year.

Cockchafer grubs* and similar insects in old pastures and grass-lands are annoying. If we are transforming a small parcel of land, for example, into a garden, it is best to fence it in and let pigs and chickens run on it for a while; their presence loosens and cleans the ground. On larger areas, an intensive reworking of the soil is essential. It especially needs harrowing to hinder the development of maggots; light and air are their enemies. If other measures do not help, "trap plants" can be employed by sowing spinach or scattering pieces of potatoes, which are later gathered up covered with these pests, thus protecting the crop planted in the field. Hence, under some circumstances, planting peas (even with the large amount of hoeing required) is advisable on old pastures.

Preparations 500 and 501 have a role in handling meadows and pastures biodynamically. Preparation 500 should be used in the autumn and once in the spring; preparation 501 should be sprayed on the green plant after there is no longer danger of night frosts (cf. chapter 5), and again a short time after the first and second cuttings. We have noted especially good results during dry periods when we applied preparation 500 immediately after the first cutting, and then sprayed 501 a week or two later.

---

* Also called maybugs or doodlebugs, these are European beetles of the genus *Melolontha*, in the family *Scarabaeidae*.

When laying out a garden plot where grass has been growing, it is advisable to peel off the sod and make it into a special compost heap. The subsoil should then be worked in a normal way, and later the compost should be returned to it. Conversion of a garden shouldn't be difficult if biodynamically treated compost and manure are present in sufficient quantities, and if the principles of crop rotation cited in chapter 5 are observed. An especially intensive application of the biodynamic measures can be made in a gardening establishment. To outline this, however, would go beyond the limits of this book; hence the details must be left to the advice of the biodynamic information centers.*

It is our usual custom today to think of farm economics in industrial and commercial terms, but such thinking assumes that capital turnover in a year should show at least a 20% profit. Otherwise, there is no "urge" to start such a "business." This is presently the attitude of those involved in modern commercial farming, especially in such enterprises as the monoculture of sugar cane, citrus trees, tobacco and in dairy farming. These are usually started with the speculative point of view that a monoculture might succeed if its returns begin within a very few—maybe 4 to 6—years. The longer it takes to reach the productive period, the poorer the economic value, because this lack of balance never allows the soil condition, humus structure, or general fertility time to improve.

The production of food is one of the most important and vital problems of humanity—hardly the most fitting field for financial speculation. The economic side of food production should be calculated on the basis of a long-time rhythm, not unlike the rhythmic advance of human evolution itself.

The result of experience shows that the proper basis for an economically self-supporting farm—i.e., the basis for the capital turnover—rests on crop "turnover," the rotation of crops. This means

---

*   See, for example, H. Koepf, *Koepf's Practical Biodynamics: Soil, Compost, Sprays, and Food Quality;* also, P. Masson, *A Biodynamic Manual: Practical Instructions for Farmers and Gardeners.*

that we cannot expect a return on capital expended in a year; its return is proportional to crop rotation. If we have a three-year crop rotation, the capital turnover takes three years to accomplish. This is in harmony with a natural biological rhythm. If we have a five-year crop rotation, the capital turnover requires five years. This gives farming a healthy economical basis. A farm, then, is really a hedge against a boom in the one direction or a depression in the other.

It has been discovered, as a corollary, that the shorter the frequency of crop rotation, the more intense the work is; a three-year turnover of capital requires more implements and a greater intensity of farm labor than a five-year turnover. For example, we must spend relatively more money for implements and for labor spent on the soil. The longer the period of crop rotation, the less intensively we have to work and, thus, the lower our costs. There is less rush, less effort, less strain.

One main crop lost in a three-year rotation period means the loss of a third of the arable land (production value). In a five-year rotation plan, the loss is only a fifth, and the economic condition is more elastic and resilient in the face of outer influences. At the same time, we observe that our efforts are more biological—we are saving the humus and soil fertility. We learn by practice, therefore, that the biologically most balanced farm is at the same time the most economically self-supporting farm. The profit is small. Under present conditions, a profit of 2 to 3% is relatively high, but it is fairly well guaranteed, and the capital is safe—provided, of course, the treatment of the soil saves the soil capital, the humus.

Recently the value of a diversified crop has been recognized in the more intelligent official circles in Europe, where it has been discovered that only on soils where the ideas presented in this book have been properly applied do we find truly healthy farms. A practical condition of society (in the past) introduced the idea of the family farm. A farmer and his family did the work and kept the farm intact and fruitful for generations. The idea of both family farmer and diversified farming seems to this author to be the only basis for a truly healthy future farm life and healthy social conditions. *But this is true only when the full effect of the humus-saving biodynamic system has been attained.*

CHAPTER 8

Forestry

We can learn a lot from forestry about various aspects of soil biology and plant associations. The same laws we see at work over long stretches of time in the forest are—compressed into a few months—applicable to farm and garden culture. The formation of immature humus is an example. Numerous leaves and evergreen needles fall to the earth and are cut off from access to air. In time, this produces sour humus. Lack of air, with otherwise satisfactory organic (but unrotted) fertilizer material, always produces sour humus. In either case, lime and ferrous salts, for example, are washed out of the upper layer and deposited in the lower. Thus, the so-called meadow ores are formed. They consist in every case in the development of a middle layer, which isolates the upper and lower soil levels. They interfere with groundwater circulation and mark the beginning of the soil "sickening" and becoming infertile.

If the hard layers are not very deep, they can, of course, be cut up with a subsoil plow (mole). Many deep-growing plant roots also penetrate this hard layer. Stinging nettle is such a plant, which helps to decrease the iron content in a soil by forming free ferric oxide (Fe_2O_3). In woodland soils, the black (or yellow) locust tree (*Robinia pseudacacia*) can similarly be of beneficial service.

The situation is more serious when the isolating layer lies deeper, as is frequently the case in areas where steppe land is in the process of forming. Despite our advocacy of intensive soil culture in farming and gardening, we cannot generally advise the cultivation of the soil in woodland areas because of the expense and meagre results. For experimental purposes the soil might be torn up occasionally with a subsurface plow to allow air into the immature humus layer. But here, too, the effort to find a natural solution of the problem ought to be emphasized.

In earlier centuries of European forest economy, family farmers would drive their pigs and cows into the woods. If this is done with discrimination and on a reasonably small scale, it can be of value, since it not only loosens the soil but also adds manure. But it is regrettable to have to say that woodland pasturing has been over-done; pigs injure the roots, and cows eat seedlings and young tree shoots. As a result, many woods have been ruined. Use of animals as a regulative measure however, tried experimentally and in a small way, is often interesting. A movable paddock containing relatively few animals—small enough to set up and move frequently from place to place, but large enough that "intensive" churning of the soil doesn't occur—can be used to loosen soil and get rid of weeds.

If farmers are able to give their pigs such an outing, it also helps to make them healthier animals. In this connection, we are reminded of the well-known strong, black pigs of Monte Cassino near Naples, which are pastured in this way. Chickens can also make themselves useful in this way. As a flock they spoil the soil, but a few in a large yard can get rid of the mossy growth on a turf, help aerate the ground, and in addition eat troublesome larvae and other pests. Chickens, as well as other animals used in this way, must have their yard moved about frequently.

An important factor in forestry is the question of monoculture versus a mixed stand of trees. With few exceptions (oak, for example), monoculture has shown itself to be harmful. In a monoculture, the roots of the trees take nutritive substances from the soil in a disproportionate way. The single crop has a one-sided effect on the soil's acidity, and the falling needles or leaves produce only an unbalanced humus. If for example, only beech leaves fall to the ground, they eventually bake together into a thick, impermeable layer without mixing with the soil. The fertilizing effect of tree leaves is frequently not fully realized.

Pounds per acre of fertilizing mass in a normal stand

Beech	3,614 lbs.
Spruce	2,232 lbs.
Pine	3,261 lbs.

Keep in mind that falling leaf masses are rich in organic (tannic) acids and in mineral acids (SO_3). Here again we see a source of substances that can help open up the fertilizing resources of the soil. One kilogram of dry leaf material, for example, contains SO_3 in grams as follows:

Fern leafage	1.57	White pine needles	0.62
Wood moss	1.10	Spruce needles	0.47
Beech leaves	0.73	Pine needles	0.35

The past hundred years have certainly proved that forest monoculture is harmful. The imbalanced soil reaction has already been mentioned. Another point against it is the swift and unhindered spread of pests and plant diseases. Then, too, a stand composed of only pine in a dry area should be especially feared in terms of forest fires, because of its high content of pitch. A mixed stand of various kinds of trees, including both evergreens and deciduous varieties, has proved itself everywhere to be the most stable biologically and the most advantageous.

We still need to consider, in this connection, the mutual influence of plants on one another. The total knowledge of all the positive and negative influences of plants and their roots on one another, as well as their mutual natural compost fertilizing, is still very small today. Yet, despite our limited knowledge, we recognize the vital significance of such interdependence for the health and biological capacities of plants and trees. For example, any other tree may be grown with the oak. Beech, too, has a beneficial effect in a mixed stand. Spruce, by contrast, is a thief; it suffers no other tree in its vicinity, but spreads out and, in time, chokes off all other trees, except when growing in the neighborhood of the birch, where it is *greatly stimulated*. We can thus study the mutually beneficial and harmful influences of plants in every climatic region. It might be possible in this way to formulate fundamental principles for the healthy practice of forestry.

Mixed-tree forests produce humus that has numerous components, and is therefore loose, crumbly, and permeable, and the spread of harmful insects is checked. In short, here we have the most beneficial biological conditions. And, especially since the fertilizing of forests is not practical, we need only allow the laws of biology to take their course.

There is a decided difference between forest and garden in terms of fertilizing, since fertilizing with organic manure is absolutely necessary for agriculture, but not for forests. Moreover, in the judgment of experts with whom we consulted in various countries, mineral fertilizing of forests has proved a failure. For quality, a definite specific rate of tree growth is needed. An increase of soluble salts in the ground induces a tree to take up more salts. Since a balanced salt concentration is always present in the plant cells, if more salts are taken up by tree or plant, more water is needed. But more water means *swifter growth,* as well as *looser, softer pulp* with larger cells; mineral fertilizing also involves a greater withdrawal of water from dry soils.

An interesting observation may be injected here. The desire for higher yields is often expressed by farmers and is frequently the force that drives them to unwise actions. They think of intensive fertilizing as the only means of attaining greater results. But it must be kept in mind that water is one of the most important elements in a plant—40 to 80% of its green mass consists of water. Increasing yield, therefore, means more water, among other things. Intensive fertilizing of plants is therefore generally possible only when enough water is available. In Central Europe, the groundwater level has been sinking for a number of years; hence, the water situation will make increased yield impossible unless the water level is regulated. The average lowering of the water table in 10 years has been about five feet, though regions are known where this sinking has gone as far as 50 feet. Furthermore, it has become clear that canalization and stream and brook diversion, as well as *building artificial backwaters and hydraulic power plants, frequently act unfavorably on the groundwater table.*

Since this issue is already becoming a burning question in a region well supplied with water, how much more important must it be in dry regions with both a scarcity of water and an intensive grain culture? Humus in the soil holds moisture, prevents early saturation, and slowly frees the moisture again, so that the ground remains damp far into the drier time of the year. Heavy soil does better in this respect than light soil. But when a heavy soil, poor in humus, does dry out, this produces the worst of all conditions. Lumps are formed as hard as cement. The table shows some percentages of water retention for various classes of soil.*

The general regulator of the water economy of a region is its stand of forests. Forests are magnets for the clouds. They hold the rainwater and represent a natural reservoir. A region poor in forests is poor in water—excluding, of course, the possibility of irrigation.

---

* In the Humphry Davy tests, soils are finely powdered, dried at 212° F., then set out in the air for an hour, after which their absorption of atmospheric moisture is determined. The same principle applies in the absorption and retention of warmth.

Infertile, purely mineral earth	3%
Coarse sand	8%
Fine sand	11%
Normal, medium field soil	13%
Fertile alluvial soil	16%
Fertile soil very rich in humus	11%

Herein lies the tragedy of China and of conditions arising in the United States (see chapter 1). An urgent, intensive afforestation of the Chinese hill chains, and of the plains of the American "dust bowl" and adjoining regions is the only salvation.*

Stuart Chase writes the following in his article "Slaves of the Flood" (*The American Magazine,* May 1937):

> Floods are unquestionably growing worse. At Memphis the gauges showed 35.6 feet in 1890, the highest water ever recorded then. In 1916, the level crept up to 43.4 feet. In the great flood of 1927, it rose to 45.8 feet; in 1937, to 50 feet. Are we receiving more rain on the average? The Weather Bureau records do not show it. The terrible droughts of 1934 and 1936 do not indicate it. No more water is coming down from the sky, but what does come down reaches the main rivers in faster time and in greater volume. Scientists have demonstrated that soil erosion is a cumulative process. It grows like compound interest. The worse erosion becomes in the hills, the more sudden and disastrous the flood in the lowlands....
>
> Before America was discovered [by Europeans], Nature has carefully laid down the top layer of soil, called humus, at the rate of something like one inch in every 500 years.... A pound of sand will absorb only a quarter pound of water; a pound of humus will absorb twice its own weight. This girdle of humus thrown over the continents of the world, seldom more than a few inches thick, is the source of all land life....
>
> Along comes the impetuous White People. They drive out the Indians who have respected Nature. They cut down the forests and clear the land for crops.... They plow the sod of the natural grasslands, or they herd too many cattle or sheep on the sod and ruin the cover by overgrazing. They drain the swamps, marshes, and ponds for croplands, often to find that the underlying soil is useless for agriculture. In arid areas, such as the Central Valley of California, they sink

---

* *Afforestation* (in contrast to *reforestation*) is the establishment of a forest or stand of trees in an area where there was no previous tree cover.

huge pumps into the artesian basins and rapidly exhaust the underground waters.

Presently, the precious layer of humus is being washed away by sheet erosion, scratched away by finger erosion, and torn away by gully erosion. Three billion tons of rich soil are carried to the oceans and Gulf of Mexico every year.... Already, more than 300,000,000 acres of good farmland have been completely devastated or seriously damaged by water erosion. The very skin of America is bleeding away to the sea. Out in the dust bowl, wind erosion has brought another 100,000,000 acres close to ruin....

Three things have held the water on the forested hillside—first, the litter itself (twigs and leaves on the forest floors) has absorbed the water. Second, the humus, soil under the litter has drunk the water like a sponge. Third, and perhaps most important of all, the litter has acted as a filter and kept the rainwater relatively clean, so that it does not muddy and clog the pores of the soil....

Now, what has happened in the cornfield? There is no litter to hold the rain. The humus, after excessive cropping, is not as deep as it is in the forest and cannot absorb as much water. Worst of all, the rain quickly makes mud as it strikes the bare earth, and a thin, water-tight film forms and seals the soil's pores, so that water cannot get through to underground reserves. Naturally, it must tumble downhill, taking topsoil as it goes and carrying silt to dams, cities, and river beds a thousand miles below.

Dr. F. B. Howe, Professor of Soils at Cornell University, measured the runoff of rain on an acre of corn land during a growing season, compared to an acre of meadow. The plowed acre showed an excess runoff of 127,000 gallons. He found the grass to be 65 times more efficient in preventing erosion and 5 times as effective in holding water on and in the land.... In some areas as much as 85% of rainfall runs off immediately....

We can hold flood crests down on any river, anywhere, if we first learn what Nature demands, and then work with her.

But afforestation is not something that can be accomplished in just a few years. The proper development of a forest requires certain environmental conditions. The previous complete stripping of relatively large forest areas has already disturbed the biological balance. When land has been stripped of all its trees, the humus layer—the moisture of which was formerly conserved because it was well covered and shaded—is suddenly exposed to wind and sun. Soil life is disturbed and vegetation hostile to the development of woodland spreads over the ground. On sandy stretches, this process is connected with the formation of infertile heath lands. Stony hillsides remain bare; in southerly regions they are gullied by erosion. Hence, careful foresters do not cut brutally into a stand of trees but pick their way—that is, each year they remove a tree only here and there, so that others may develop and get more light, while the ground remains shaded. Only in this way can a stand of trees always remain young. If at felling time we go through a forest managed in this way, we would be amazed at the number of trees lying on the ground while the woods still stand.

Woodland trees are accustomed by nature to growing up in the shade of others. Even the trees that later need more light prefer shade in their earlier stages. Foresters must see to it that the natural spread of seeds in a forest is helped. Birds also help here. In every case, fenced-in, protective enclosures should be put up so that young plants are not devoured by animals. Within those enclosures, the ground can be loosened somewhat and given small doses of biodynamic compost. Then we will see especially effective rooting of woodland seeds in that area. Keeping a close watch on these natural processes seems to be the primary task of foresters. They must participate as friends and helpers in the growth and in the maintenance of a natural condition of balance.

The use of biodynamic compost in planting holes has also proved valuable in nurseries of young trees, which should, insofar as possible, always be laid out in a clearing between mature trees—that is, within the atmosphere of the woods.

The black locust (*Robinia pseudacacia*) has shown itself valuable as a protective plant for afforestation and interplanting. It is a legume, aiding nitrogen fixation. It has deep-growing roots, which give it a connection with the lower soil levels and the possibility of maintaining itself in dry times and in arid regions. It is relatively without any requirements. A good procedure is to set out protective strips of locust, behind which the rest of the afforestation can be carried further. The well-known German surgeon Privy Councillor Professor Bier has in two decades developed a model forest on dry sand in the eastern part of the Province of Brandenburg. He has done this with the help of protective locust trees and elder hedges and by supporting the natural seeding processes by the use of other assisting plants.

It should be an axiom of forestry, that a forest should never be completely cut down. The art of forestry lies in a properly timed felling program, and the proper spacing of new growth on the land. On hills or levels where there are no woods, it is not possible simply to plant young trees. They will be unable to find any "woods earth" there, and in most cases the trees will die off again after a few years. Here one must study nature's method of afforestation. On a dry plain, certain weeds are carried by the wind to the moistest places, most protected from the wind. Broom shrub, for example, grows on sandy soil. We see here, again, the important influence of legumes. Under the protection of broom, and with the help of its soil-improving qualities (cf. chapter 10), other plants can develop small shrubs, perhaps even shrub oak and locust for instance. Here the human being must take a hand in guiding the natural forces. Locust and broom if held in check, supply the necessary environment for tree growth. If allowed to run wild, they act like weeds, choking out everything else. It is impossible here to give the names of suitable protective plants for every climate and soil. The intention is only to demonstrate a principle.

A whole series of shrubs can grow within the aforementioned shelter. Hazelnut and elder in particular provide shade for starting young trees. Their fallen leaves and root activity produce the humus in which woodland trees like to grow. We must slowly build up a forest in this way, beginning with ground cover, followed by larger shrubs as protective plants, and finally planting the desired trees. Such a program needs 3 to 5 years of preparation, but obviously gets more quickly to its goal than by direct planting of trees, which die off after a couple of years. When dealing with afforestation of dry regions, it is wise to begin in the section where the best conditions for rain, moisture, or dew still prevail, and from there slowly extend the edge of the woods into the poorer sections.

A farsighted grasp of the situation is essential here. One must begin at a point that is still touched by moist winds to create a stand of moisture gatherers, and from there move out slowly to the dry areas. In North America, this means afforestation of the Rocky Mountains where possibilities of forest still exist, and then proceeding outward from the edges of any still-remaining wooded areas. Afforestation could be started along the upper reaches of the great streams.

This plan would work much more quickly than planting trees on isolated spots. In North America, as in Italy, we have witnessed one of the greatest errors that can be made by human beings—namely, complete deforestation by logging mountains and hilltops. The water that collects on the hilltops is thus eliminated, and then the hills themselves become eroded and bare. The water flows into the valley, forming marshes, and thus both hills and valleys become useless. A properly wooded hilltop, by contrast, creates reservoirs of water and fruitful valleys.

Completely clearing fallen leaves from woods and forests has always proved detrimental. The fertilizer that the woods produce themselves should be left to them. Leaving only the small twigs and a portion of the fallen leaves as a soil cover, however, has been found satisfactory. The main point is that the woods should not look as though they had been swept clean with a broom; for all the life in the soil and underbrush must be protected.

In a wild forest, where everything grows without help or hindrance, the trees mutually choke themselves out. Woods, naturally, need a regulating hand, for otherwise an unhealthy, disproportionate condition will develop with time.

CHAPTER 9

Market Gardening

The reader might ask why an outline of the biological laws of the forest should precede the chapter on gardening. There is a simple reason for this. The basic laws of the forest—for example, constant ground cover, mutual help of various plants, protective planting strips, and compost fertilizing as naturally as possible—apply similarly to gardening. If all that can be learned about the biology of the forest, about perennial forest plants, is compressed into a single year, we discover that the knowledge thus gained of forest conditions makes it possible to establish a truly intensive garden culture.

For every soil that tends to dry out—whether in a garden or farm in a temperate or tropical climate—mulching and shading create conditions in which healthy fermentation can develop and losses of humus are avoided. For intensive gardening, it is especially necessary to pay attention to these matters.

A garden, too, must be protected against drying winds. The hedging needed for this also keeps the garden warmer, which in turn brings the vegetables to ripening several days earlier, which means greater profit. Such an arrangement includes a protective strip of hedge plants or trees around the garden. A hedge 6 or 7 feet high keeps wind off the ground for a distance of more than 300 feet and raises the soil temperature in spring from 1° to 2°C. The south side and the side that receives the prevailing winds should especially be considered. Where there is heavy wind pressure, it may be that poplar or wild cherry are the only possible as windbreaks. For wet soils, an alder tree is recommended because of its nitrogen nodules and the drainage effect of its roots. Alder is especially suited along ditches and canals. Other good shrubs and trees for hedging are hazelnut, blackthorn, or birch. Of course, this is only a hint of the possibilities.

In a warm, misty valley, there should be protection against the cold winter winds. At the same time, air circulation must be provided. This is especially necessary where fruit trees are planted. Fruit trees require air currents for the ripening process. With the exceptions of espalier varieties, fruit trees should not be protected by a wall or by a windbreak hedge that is too thick. Lichens and moss are signs of stagnant moisture in the air.

On a larger tract of land, protective hedges should be used to separate individual fields, as well as to enclose the entire area. If the main hedge is a line of shrubs or trees, the inner rows can well be fruit trees, or berry bushes. One's plans should always carefully consider alternations in planting rather than monocultures. The more varied the planting, the better is its biological effect. The same is true here as it is for woods. A large area of only apple trees or only currants is much more susceptible to attacks by pests than areas broken up with varied plantings. It might be argued that this complicates harvesting—perhaps, but isn't it also complicated and troublesome to spray copper sulphate, arsenic, and lead preparations, *as well as detrimental to health?* By means of other preventive measures still to be discussed, this sort of spraying can be avoided. The labor required to carry out one method or another is found to be about equal when considering the work in each case as a whole.

The area planted in vegetables should also be broken up by rows of high-growing and low-growing plants, which give shade in times of drought, help to bring about healthy soil fermentation, and provide windbreaks. To obtain the beneficial effect of legumes, pole beans, pole peas, and so on are useful. Sweet corn can also be used as an inner hedging plant. In the shelter of these rows, all the lines of low-growing plants can be set out. The most intensive planting system of the Chinese, which often employs as many as six varieties of adjacent plants, has produced good results for thousands of years. In this case, special ridges are made in the ground. One kind of plant is set on the ridge, which grows swiftly, with another kind of plant between the ridges. Being shaded and kept moist, it grows unhindered by dryness. As this latter plant grows higher, the one on the ridge is ready to harvest, and then the plant growing between can in turn shelter the ridge.

From forestry, we learn the value of a mixed culture and the disadvantages of monoculture; from agriculture, we learn the need for a healthy crop rotation in gardening. Here, too, there are plants that do not harm the soil—for example, any legume. There are also plants that exhaust the soil—for example, nearly the whole cabbage family, especially cauliflower, as well as celery, cucumbers, and leeks. Carrots, salsify, beets, radishes, small turnips, onions, lettuce varieties are also among the plants that take little from the soil.

The point here is to begin by working out the most beneficial crop rotations. In place of manured and hoed crops in farming, in a garden we have early potatoes and cabbages, which also need manure and are among the heaviest consumers of soil substances. Especially important for good results is that early potatoes or corn can be planted in rows with bean rows between.

After these, as the next stage of rotation, crops with fewer soil requirements can be employed, including field salad, spinach, kale, kohlrabi, head lettuce, endive, and Brussels sprouts, provided these latter are not planted after cabbage but, for example, after potatoes. Plants that do not exhaust soil so much are particularly suitable as preceding crops. These include spinach, early root vegetables, field salad, lettuce, and kohlrabi.

However, it has been shown that not only plant rotation but also the juxtaposition of plants have an important effect. We need only try planting tomatoes and kohlrabi in a mixed culture to be convinced of the poor influence of kohlrabi on the tomatoes; or we might plant two to three rows of radishes alongside two or three rows of garden cress, plus another test planting of radishes with chervil growing alongside. Then, if a "control" planting without border plants is made, we will notice a clear difference in the taste of the radishes.

Radishes from control bed with no border plants — poor flavor
Radishes with chervil as border plants — sharp flavor
Radishes with garden cress as border plants — excellent flavor

If we want tender radishes in summer, they do particularly well when planted among leaf lettuce This is discussed from the scientific perspective in chapter 10. Here, we wish only to give what has been observed and tested in many localities.

For many years, this author has directed a commercial vegetable garden of about 21 acres. Monoculture had for many years been practiced in this establishment. There were large plantings of only spinach, only beans, only cabbage, and so on. Based on our experience, we gave up that method and today follow completely and profitably a mixed system. Here, too, we must proceed with common sense and not in an abstract way. For example, spinach, corn, and salad or leaf lettuce should not be placed in alternate rows with other plants, but in beds. Anything that is transplanted does well in single rows. When beds are planted, they should be alternated just as the rows are—that is, always with legume beds between. If desired, carrots and peas can be planted in alternate beds, with radishes or leaf lettuce in rows among the carrots. Mutually beneficial when grown alongside one another are leeks and celery; carrots and peas; early potatoes and corn; cucumbers and beans; cucumbers and corn; kohlrabi and beets; onions and beets; early potatoes and beans; tomatoes and parsley. Further study of such relationships will certainly continue to produce much useful data.

Harmful combinations are tomatoes and kohlrabi; tomatoes and fennel; and fennel and bush beans. Any aromatic herbs and pot herbs make good border plants. Other beneficial combinations to be noted are turnips and peas; bush beans and celery; cucumbers and peas; cucumbers and bush beans. Where heavy soil-consuming plants have been grown with manure, they can be followed well by bush beans, but not by peas. It is more difficult for the latter to stand the direct aftereffect of manure.

An example of good combinations would be: two rows of celery, alternating with two rows of leeks. Now and again, using a wider space than usual between the rows, bush beans may be planted between the celery rows already mentioned, but only two beans to the hole. In very good soil onions and early lettuce can be put together. The lettuce is harvested and the onions spread out. Although onions grow poorly in sand, they manage to get along in it quite well if chamomile has been sown thinly in between.

When sowing legumes—peas in the garden or alfalfa in the field—it is preferable that they are planted in double rows. These plants will give one another mutual aid in their growth and will still have on the outside, between them and the next double row, enough room for development.

It is not our purpose here to present a gardening textbook. Those who wish to use these suggestions should already be experienced gardeners. These references are intended instead to point out problems that have not been previously observed and studied thoroughly enough.

The question of manuring is particularly important where garden land is used intensively. Thus far, we have treated the aspects of this issue that deal with soil conservation and that avoid methods resulting in soil impoverishment.

The care and treatment of manure has been discussed in chapter 5, but for gardening purposes there should be an intensification of this treatment. We only outline the principle here. What is done with it must be left to the individual. As the first instance, raw, strong, and intensive manuring (especially the use of liquid manure) forces the plants to heavy growth, producing thick, green leaves. When these vegetables are cooked, the odor in the kitchen tells us the sort of manure used in their culture. Such variations of odor arising from the differences in the manures used are very obvious as already stated in the case of cauliflower, for example.

In such raw, immature manure, the strong-smelling products of partially disintegrating albumin and nitrogen compounds have not been completely consumed, but have been taken up by the plants directly from the soil. Aside from their odor, these compounds also have certain incidental and harmful effects on living things in the soil and on the plants themselves (promoting susceptibility to fungus diseases), as well as, when eaten, on the human organism, including digestive disturbances, heartburn, bloating, and, if the process goes far enough, even intestinal worms. Such worms can be seen in the muck after the night soil fields near cities are drained. Many stomach and intestinal disturbances are cured without treatment when the use of vegetables fertilized in this way is discontinued.

Two aspects of the question to be emphasized here are: 1) intensive fertilizing through the greatest possible conservation of manure values (cf. chapter 5); 2) the needs of hygiene and health that demand the best taste possible and the very best quality in the products of the garden. The more finely rotted, the more completely transformed into odorless humus a manure is, and the more is it suited for gardening. The more this manure is mixed with compost, the finer and more aromatic are the vegetables.

Thus, we see the need of manuring with prepared, well-rotted stable manure as the basis for early potatoes and other plants with heavy soil requirements, and when we wish to grow an especially delicate crop, special manure should be prepared for it. The simplest is the mixing of half-rotted manure with half-rotted compost. Both piles may be located near one another and in turning them, the two piles may be combined in alternate layers (at which time the preparations may then be again inserted), or fresh manure may be added to three-month-old compost. There are many possible combinations. This composted manure (which alone might also be mixed with alternate layers of soil) or mixed compost is suitable for all finer plant cultures, greenhouses, and so on. Once it has completely turned to soil, it can be added in small amounts directly into the plant hole or the furrow prepared for any kind of plant. This procedure makes possible complete utilization of it by the plant. This high-quality manure compost is just as suitable for potted flowering plants as it is for fine vegetables.

Normal, healthy tomato plant from a greenhouse
treated biodynamically for seven years;
height of the whole plant: 6½ feet

In practice, we prepare the seed bed first and then use biodynamic preparation 500 directly on the soil, into the open furrow or planting hole, adding in some compost, and then sow or plant. When transplanting, it is best to give the roots a bath in a dilute solution of preparation 500.

For cucumbers, a compost with a heavy proportion of manure is advisable. Like tomatoes, these may be planted on high ridges or on little mounds of earth. A handful of compost earth is then put in every seed or plant hole. It should be added that cucumbers and tomatoes do poorly together. A suggestion by Rudolf Steiner in connection with tomato culture has proved particularly valuable in practice. In general it is better not to grow a plant in compost made from its own remains (this is especially important for cucumbers and also for cauliflower and other cabbage varieties), but Dr. Steiner called especial attention to an exception to this rule in the case of tomatoes, which are particularly at home when grown in its own compost. All the tomato refuse, leaves, and stems are set up in layers with earth according to the familiar composting technique. The compost is thus ripe in time for the next season and is put into the planting hole.

For greenhouse culture of tomatoes, it is especially important to have good ventilation and dry air. Moist air under glass promotes the development of the fungus pests. Underground irrigation has been found valuable here. Pouring water on the plants always adds dampness to the air. Underground watering allows plants to have the necessary moisture while keeping the air dry. Perforated pipes are laid at a depth of about 10 inches at a slight incline and parallel to the plant rows. These are then provided with water from a central channel. The amount is regulated so that the moisture is just visible at the surface of the ground. This procedure will not work in porous, purely sandy soils because of the great water loss such ground would entail. With this exception, the method has proved valuable everywhere, and above all it has been helpful in the fight against fungus pests. With such irrigation and the use of tomato compost and the other biodynamic procedures, for seven years now we have been able to get full yields with no plant diseases from annually repeated tomato culture in a greenhouse of 1,000 square yards.

In the first year there was still some blight and leaf curl, later there was none. It was found necessary to grow our own seed. We observed in our gardens that only with our own seed were we in a position to keep the plants permanently healthy.

The tomatoes we harvested were 87 percent A grade (the most salable, commercial export size), 7 percent of B grade, which were larger, and 6% of C grade, or smaller tomatoes. This was the figure year after year. It was observed that the fruit clusters developed evenly on the plants and trellises. An even, healthy growth was in fact one of the most important results of this method of using fine-quality compost.

These special composts are capable of virtually every possible variation. To describe them all here would take up too much space, but it should be emphasized again that cauliflower compost should not be used on cauliflower plants, or cucumber compost on cucumbers. Composted pig manure is especially good for leeks and celery. Composted chicken and pigeon manure can both be put to good use, for example, in horticulture, and in all situations where a strong forcing effect is needed.

We come now to a chapter of gardening that, we regret to say, has become very important—the subject of pests. Instead of using poisonous copper, lead, and arsenic preparations, the pests can be attacked biologically. For this purpose, the life rhythm of the plant and the pest that has attacked it must be studied. The presence of plant lice on broad beans is an instructive example. They attack the beans at a certain point of their growth—that is, from the formation of the fourth leaf group to the development of the eighth. If weather conditions are beneficial, the plants eventually get over the attack. If the contrary is true, they turn black and die off. Careful observation here can teach us a great deal. The attack of plant lice becomes especially strong where there is insufficient air circulation around the beans; this occurs when they are planted too thickly. It is also strong when cold or drought suddenly checks the growth of the plants. Beans should be planted in series of a few rows, and then their growth should be closely observed. Once the danger point is reached, they should be sprayed with preparation 501, which stimulates plant assimilation and upward growth. This helps the plant grow swiftly past the critical point and more quickly reach the later stage, when the sap develops a taste no longer agreeable to the plant lice.*

Another example is aphids on fruit trees. Rudolf Steiner suggests planting nasturtium among the trees. We carried this suggestion further, making not only the plantings but carefully painting and spraying the trees with an extract of nasturtium. Such plantings are effective, because the nasturtium contains a strong aromatic substance that also penetrates the ground through the roots. The trees can take this up through their roots and bring it into their sap stream, making changes in it that are imperceptible to us but are evident to the fine organs of smell and taste of the insects. Result: The aphids disappear. Also, of course, we should not forget that proper treatment of fruit trees is of great importance in helping to restore their powers of resistance to pests.

Another observation may be made concerning flea beetles. These are driven away by shade. They prefer a crusty soil, the surface of which, swept by the wind, becomes dry and impermeable. Hence, we should work to develop a crumbly soil with strong capillary action, and to make a shade by means of mixed crops and inter-crops. We should also mulch the soil between the rows with partially rotted leaf compost. The flea beetle shuns tomatoes and

---

* Tests in the United States, utilizing the same procedure to help bush beans against the Mexican bean beetle, have shown most satisfactory results.

wormwood. It is therefore advisable to plant these at random, here and there, between cabbage or radish plants. Even the spreading of the trimmed-off shoots of tomatoes can be helpful.

Another pest we have to fight is clubroot. Its development is promoted by the use of unrotted, raw manure; or of unmixed manure, such as pig dung only, or goat dung only, or of raw, uncultivated, or poorly worked soil, or of too little manure. Large amounts of ripe, mild compost and later also compost from herbs and vegetable plant refuse put into the planting hole can be helpful in overcoming this condition. To make possible the early planting of cabbage seedlings, we can use some leaf or straw compost as a moisture-holding mulch around the plants.

Since the cabbage butterfly is repelled by hemp, tomatoes, rosemary, sage, or peppermint, it would be wise to use such plants as protective inter-crops. The asparagus beetle is repelled by tomatoes, the mole cricket by hemp, birds by a salt herring hung from a pole. Birds are also very skittish and shy when confronted with the decaying body of a bird of the same species hanging from a pole. Scarecrows fulfill their purpose only when they are moved about frequently. Strips of any glittering material, hung so they are in continuous movement, make an excellent scarecrow.

To combat blow flies (*Calliphoridae*), blossoming trees must be given a highly diluted liquid spray of slaked lime. It is often also advisable to sow or scatter the seed of "attracting plants" to entice insects from plants needing protection. In this way we can use lettuce, spinach, and potatoes to combat cockchafer larvae, strawberry weevil, woodlouse, ear worm. There are many possibilities for helping oneself in such natural ways. For instance, to catch snails, empty half skins of oranges or grapefruits may be laid between the beds with the open side down; after a certain time, a whole collection of these pests will be found under them.

For the care of fruit trees there are some special rules. Biologically, the fruit tree stands between the woodland tree and the cultivated field plant. It needs care, but no intensive fertilizing. It has in any case a longer cycle of growth than the annual or biennial. The four chief causes for trouble in an orchard are: 1) too strong fertilizing, especially if this has been done with fresh, raw manure; 2) too thick a stand of tree, which permits too little light and too little movement of air; 3) the wrong tree stock, for the tree in question; 4) and finally, the use of a variety of fruit tree on a soil and in a climate to which it is not suited. These four points must be considered by those who aim to improve the general health of their orchards. Obviously, results cannot be obtained by working contrary to nature. If the wrong rules of procedure have been followed and the constitution of the tree has been weakened, the impossible should not be expected.

One thing should be kept in mind above all: Only the best, rotted manure and compost helps the tree. Only when this is used does the fruit ripen well, stay on the tree instead of falling too soon, and retain good keeping quality. Too intensive fertilizing or use of raw manure is sure to bring the opposite to pass. Here the relationship with the forest tree may be seen. Around the tree a firm, matted, mossy, crusty layer of soil should never be allowed to form. The roots of trees like those of other plants, prefer a loose, airy soil. If the tree stands in a meadow the soil must be loosened from time to time by digging up under the overhang of the widest limbs and branches, and then there should be a light fertilizing of the soil with prepared compost and preparation 500. It is useless to put fertilizer close to the trunk because here there are no fibrous roots to take up nourishment. Where the field is to be used exclusively as an orchard, it has been found valuable to keep the ground open for two years, digging it up as often as necessary, then covering it afterward for two years with a legume, such as clover. This latter can be used also as fodder or composted. In this way provision is made for natural nitrogen fertilizing and for a loosening of the soil. In this way the fruit tree is not sated with N_2, a substance to which it is extremely sensitive. Among our special measures against fungus diseases we include spraying with preparation 508 (*Equisetum arvense*). We use this in fruit growing by painting it on the trunk and lower limbs of the tree and by spraying the crown. This must be repeated as often as necessary, and is most effective when begun early as a prophylactic treatment.

Special benefit has also been found from painting the trunk and spraying the crown with a mixture of one-third clay, one-third cow manure, one-third sand, plus as much water as needed to make the mixture thin enough to paint or spray. In the water either 500 or 508 may be used. This is applied in the autumn and repeated, if necessary, before sprouting time in spring. It stays for many months on the trees and helps the formation of a healthy, dense, enclosing bark. It stimulates the cambium layer, heals wounds and stops bleeding of sap, and in general it has shown itself to be of great benefit to the health of trees. Cankers, cleanly cut out and painted with this mixture, heal with smooth edges. The use of tanglefoot as insect traps should never be ignored in the case of trees. In early spring, the bark of the tree should be thoroughly cleaned for several inches about three feet above the soil. The cleaned part of the stem should then be covered with the liquid glue material, and on this should be laid a clean strip of cotton cloth (linen), also covered on the outside with tanglefoot glue. This is the best trap for any beetles migrating from their winter rest in the soil to a home in the tree.

The following experiment has been tried with the clay, cow manure, and sand paint previously mentioned: groups of flowerpots were painted with a) a copper and lime mixture; b) a carbolic acid solution; c) the above clay, manure, and sand paste; or d) nothing, as a control. Tradescantia shoots of equal size and age were set in a mixture of sand and humus in the pots. After a period of time, the plants were dug up, and the development of the root weight was determined. The clay, manure, and sand paste (c) had an especially strong effect on the development of the root weight.

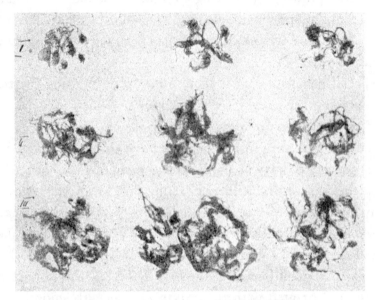

The roots in an experiment like the first

1. *With copper and lime mixture (top)*
2. *Without any special treatment (middle)*
3. *With the clay, manure, and sand mixture (bottom)*

AVERAGE FIGURES FOR 16 PLANTS IN EACH CASE

Weight of the shoots at planting	Weight of the roots in grams at conclusion	Weight of green plant parts
a. 2.87	3.51	35.25
b. 2.58	3.11	43.17
c. 3.0	6.62	46.8
d. 3.18	4.11	49.12
Second experiment		
a. 5.02	6.35	17.23
b. 4.72	8.73	30.8
c. 5.05	4.97	28.62

It is advisable to wash down the trees with preparation 508 before using the mixture, having first brushed any fungi and algae off the trunk. The argument one hears is, *That makes more work!*...as if extensive spraying with all sorts of other material in recent years doesn't also make work. The work of the two procedures balances fairly well; it is about the same in either case, but with this method we have the satisfaction of knowing we are working in a healthy, natural way.

The same treatment has also proved its value for grape culture. Grapes, in addition, need an intensive spraying with preparation 508 during the critical time of the early summer. With fruit trees and berry bushes, a light compost fertilizing in the late summer is advisable, after the fruit is well-formed and while the leaf masses are still green. The set of the fruit for the next year is helped by this treatment.

For strawberries this is done in August, after the harvest, so that no harm is done to the fruit. This fertilizing must not be strong enough to have a forcing effect.

To complete our picture of the biodynamic method, a whole series of detailed procedures still remains to be outlined, but it is not possible to complete such a list in the space of this book. It is better to leave such questions to be resolved as they arise through the advice of biodynamic information centers and publications.

CHAPTER 10

The Dynamic Activity of Plant Life

The nourishment of a plant consists in the assimilation of salts, of water, and of carbonic acid taken from the air. Its total mass is composed to the extent of ninety percent and more of water. Only from two to five percent of its mass comes from salts taken from the soil. Thus, *it is a surprising fact that the plant receives an important part of its nourishment from the air,* which we can neither fertilize nor influence in any way—while *only a relatively small amount is received from the soil,* which we *can* and *do* influence and fertilize.

Experiments, for example, which consisted of increasing the CO_2 content of the air in greenhouses, have had no practical significance. We have little influence on the volume of nutritive material coming to the plant from the air. Only at one point can the absorption of carbonic acid be strengthened: by increasing the so-called soil carbonic acid. Considerable research has shown this soil carbonic acid to be an especially active and beneficial agent in plant growth. Soil carbonic acid is produced by the microorganisms in the soil, and thus depends on the degree of humus and microscopic life in the soil itself. Here we can influence the plant's nourishment coming from the air in connection with one of the most essential nutritive materials.

These carbonic acid salts may be taken up directly by the roots if the salts are present in a soluble, that is, in an absorbable form. It is, therefore, to the farmer's advantage to bring the soil into the proper condition. This is accomplished more easily when there is considerable water in the ground. We must, therefore, look after the soil's water supply in dry regions. Humus and a crumbly structure are important factors in this connection, as we have frequently mentioned. They help in the process of weathering, and making available the mineral substances in the ground. This process is also advanced by the plant roots themselves, in part mechanically and in part by the effect of their own secretions.* These plant properties must definitely be taken advantage of in rational agriculture. We define any treatment that enlivens the soil, making it richer in humus and opening it up and loosening it as "biologically effective." Ways to attain these ends are described in this book.

Much is said about the primary nutritive materials—potassium, nitrogen, phosphoric acid, and calcium. According to the law of nutritive substances, the volume of these materials taken out of the soil by a harvest must be replaced by fertilizing. This law—in itself correct, logical, and confirmed by laboratory research—is subject to continuous variations in natural, outdoor conditions. In practice, we do not know exactly which substances are removed and in what form; nor, on the other hand, do we know which are replaced by weathering in the soil and being carried back in atmospheric dust and water. Nor do we know whether a specific method of working the soil cuts off the access of such materials or brings them into the constant cycle of substances used and replaced. Even when we fertilize with a substance, we might disturb the balance and cause deficiencies in other salts.

We have mentioned the rhythmic variations in the solubility of phosphoric acid compounds. Briefly, in them we have a series of biological phenomena, the course of which is difficult to control, but is as important for nourishing the plant as purely quantitative "replacement." Conditions of balance or imbalance that must be considered arise continuously. For example, the nitrogen problem—it is plainly

---

*   Cf. ch. 6 for a discussion of the effect of various plants on making soil materials available.

evident that there is an antagonism between the nitrogen produced by bacteria and the nitrogen coming from the use of mineral fertilizers. Legumes that are forced with artificially applied nitrogen do not develop bacteria. Clover on meadows fertilized with ammonium sulphate disappears. Fresh liquid manure (with its free ammonia) has the same effect on clover as the ammonium sulphate.

An experiment carried out in Holland shows the following results: On an experimental field of 6 groups of 10 parcels each, 10 received no nitrogen before the first cutting; 10 were given Chilean nitrate; 10 calcium nitrate; 10 ammonium nitrate; 10 calcium ammonium nitrate; 10 ammonium sulphate. The application consisted of 40 pounds of nitrogen per acre. Every year, mowing was done early—that is, at the best possible time for mowing, while the fields were pastured for the rest of the season. Clover growth on the untreated parcel proved to be the strongest.

Analysis of the hay showed the following:

	protein %	lime %
Parcel without nitrogen	12.9	1.33
Parcel with Chilean nitrate	8.9	0.75
Parcel with calcium nitrate	8.5	
Parcel with ammonium nitrate	8.7	
Parcel with calcium ammonium nitrate	8.7	
Parcel with sulfate of ammonia	9.2	

The hay yield on the unfertilized parcel was quantitatively smaller (because of its high protein content), yet in spite of this small yield of hay the total protein yield was greater. Frequent nitrogen fertilizing drives clover out of a field; late mowing and irregular grazing does the same.

Interesting also are figures of experiments on the development of weeds in meadows fertilized with nitrogen fertilizer in pounds per acre:

Nitrogen fertilizer in increasing doses:

0	60	90	120	150

Amount of Crowfoot (Ranunculus acer) in clover fields:

0.5	1.5	3	4	5

Clover harvested:

7	4	4	3	2

This means that clover parcels without mineral nitrogen harvested 3.5 times more clover than the clover parcels fertilized with 150 pounds of pure mineral nitrogen.* We see clearly from this example that the effect of nitrogen fertilization is not only quantitative, but that it also affects the whole biological process.

In this connection, we can include another important observation. The mineral substances in the soil are in a state of equilibrium. When, however, a soluble material is strongly preponderant, then other materials are driven out of the solution, precipitated, and so on. It is well known that heavy potassium fertilization precipitates soil magnesium. Research has shown that, on most agricultural soil, there are continual disease conditions of rye and oats, and analysis has shown that these were caused by a lack of magnesium.

Prof. Schnitt's research leads to the view that, where there is strong acidity, magnesium is so vigorously absorbed by the soil that it is no longer in a condition available to the plant. The precipitation of phosphoric acid in sour soils and its fixation by heavy soluble iron and aluminum phosphate seem to be hindered by the presence of magnesium. It is suggested that the problem of maintaining the magnesium content of the soil is difficult only because there is such a variety of factors that work together. For instance, the activity of magnesium is decisively important for building chlorophyll, the development of protein, and the utilization of phosphoric acid in the plant.

---

*   Cf. ch. 12. Strong mineral applications of nitrogen lower the protein content, according to Prof. Boas of Munich.

The various factors mentioned include the condition of the soil, its magnesium content, its acid or alkaline reaction, its content of calcium and other salts, humus, and the effect of the plants growing on it. According to Scharrer,

> The absolute magnesium content of the soil is of minor importance when the question arises as to whether there is need for magnesium fertilizing. Much more important is knowing the degrees of magnesium tolerance, which might be so low in the case of sour soils that they eliminate any added magnesium by precipitation to such a degree that the added magnesium produces no effect. Because of the strong effect of magnesium, small amounts are sufficient to produce large results.*

We learn that the important thing is not the absolute volume, but soil condition. Conditions can arise that we can liken to a thirsty horse tied to a post near a spring with a halter that's too short.

Recent developments in plant physiology teach us that, in addition to the coarser substances present in plants, there is a whole series of ingredients—often present only in very minute quantities—that are absolutely essential to plant life. Such substances were not considered in the fertilizing, and yet they are present. The role of boron is familiar. Legumes develop root nodules only when there are traces of boron present. *Solanaceae* (the nightshades) such as tomato and tobacco, as well as chenopods (goosefoots), grow healthily only where there are traces of boron. In the case of the sugar beet, rotting core can be cured by spreading 4 to 13 pounds of borax per acre over the field.

The question of copper is also illuminating. Copper sulphate, in a dilution of 1:1,000,000,000, injures algae (spirogyra); in a dilution of 1:700,000,000, it hinders the development of wheat sprouts; with 1:800,000, it stops their growth. In soil, it is still a

---

* Prof. K. Scharrer, München, *Bedeutung seltener Elemente für die Landwirtschaft, Mitteilungen für die Landwirtschaft*, vol. 2, 1937.

strong poison to bacteria at a dilution of 1:100,000. Nevertheless various crops contain large amounts of it—oats up to 0.9%; barley, wheat, and rye up to 0.01%; potatoes 4 milligrams per kilogram; hay from 6 to 12 milligrams per kilo; lentils 0.015%; broad beans (*Vicia faba*) 0.03%; peas 0.01%; soybeans 0.01% of the ash weight. It is found in oranges, European elder blossoms (*Sambucus nigra*), watermelon, squash seeds, black mustard, maize (corn), pine (wood and bark), German iris, and other plants. This copper is found in these plants without the soil ever having been treated with it, even when it is scarcely possible to find a trace through soil analysis. *Here we recognize the capacity of the plant to gather the most minute quantities of certain elements and extract them from their environment.*

There are also plants that accumulate lead, such as *Festuca duriuscula* (containing up to 12.25% of Pb_2O_5, found in ash) and *Randia dumetorum*. Titanium is found in the wood of apple and pear trees, and manganese is of significance for the grapevine. Iron is stored up in many plants up to high percentages in the case of the acacia cebil, or the pine.

Hugo de Vries, in his classic *Leerboek der Plantenfysiologie* (Haarlem, 1906), presents analyses of aquatic plants that make these characteristics clear. He says that "the chemical combination definitely does not conform to that of the soil or the water in which the plant grows. And sometimes variations found in two plants growing close together are very great" (p. 201). Thus, for example, the analysis of certain plants growing in the same ditch shows the most varied results (see table, page 118).*

The capacity for absorbing such substances in highest dilution and then accumulating them belongs in common to all the kingdoms of nature. A whole series of substances, especially the heavy metals are present in the human body also. If these are present in

---

* Ida Noddack and Walter Noddack, "Herkunftsuntersuchungen," *Angewandte Chemie,* no. 37, Sept. 1934.

The percentage of various minerals found in plant ashes

	Chara foetida I	Chara foetida II	Hottonia palustris	Stratiotes aloides	% of substance in water
Potassium	0.46	0.23	8.34	30.82	0.00054
Sodium	0.18	0.12	3.18	1.21	—
Calcium	54.73	54.84	21.29	10.73	0.00533
Magnesium	0.57	0.79	3.94	14.35	0.00112
Phosph. acid	0.31	0.16	2.88	2.87	0.00006
Carbon. acid	41.60	42.86	21.29	30.37	0.00506
Silic. acid	0.70	0.33	18.64	1.81	traces

our food, only the slightest traces of them can be discovered or none at all. The new, refined techniques of chemical analysis have only lately given us the possibility of investigating these things. I am reminded in this connection for example of the work of Dr. Ida Noddack of Berlin.* Refined spectral analysis shows that one can speak of the universal presence of any substance, and in general of every substance, including thus also the heavy metals in a dilution of from 10^{-6} to 10^{-9}. Certain plant, animal, and human organs can accumulate these substances in measurable amounts. It remains only for research to discover how this occurs.

Here is a task the solving of which would lead to an appreciable enrichment of our knowledge of the phenomena of life. Hardly anything in nature is present without reason and purpose. Thus, perhaps it might be possible even today to say hypothetically that these minute quantities of substance in the plant world have a guiding and directing task to perform in regard to the various biological processes that occur. It is already possible from the study of the substances existing in large volume in the plants to see that their presence gives direction to plant growth. The author is thinking here, for example, of the influence on starch formation resulting from the absorption of potassium by the plant, on

---

*   Ibid.

protein formation resulting from the absorption of phosphoric acid, and on chlorophyll formation resulting from the presence of iron, in spite of the fact that in the latter case iron is not contained in the chlorophyll molecule but exists only in the environment. It has been shown that not only a lack, but also an overabundance of substances may be harmful here.

The presence of minute quantities of this or that substance has apparently more a functional than a nutritive significance. The total functioning of the organism is in many respects dependent on the status of these finely atomized substances. The health of the plant is influenced by them.

However, modern research goes a step further. It now recognizes "chemical effects at a distance." Dr. Ried in Vienna, Prof. Stoklasa in Prague, and Councillor von Brehmer in Berlin, by means of various methods of research, obtained the same result, showing that the presence of mineral substances, even when removed to a certain distance, can considerably influence the growth of plants.

Brehmer* reports that potassium in the vicinity of potatoes—separated by a space of air from the containers in which the plants are growing—was able to increase the growth and the potassium content of the potatoes. Stoklasa** shows that potassium (in sealed test tubes hung over growing plants) alters the rate of growth of the plants. We would add that we have been able to make this experiment, the validity of which we are able with corresponding results. Ried*** shows that the presence of potassium and other salts in the vicinity of animals can have a far-reaching influence on their growth and, above all, on their reproduction. In addition, there are numerous experiments that demonstrate the influence of irradiated and non-irradiated metals in their effects

---

*   Dr. A. Heisler, *Ärztliche Rundschau*, 1932, no. 1.

**   *Kosmos*, no. 12, Mittlg, 1933.

***   Dr. O. Ried, "Biologische Wirkung photechischer Substanzen," *Wiener Medizinische Wochenschrift*, no. 38, 1931.

at a distance on the development of bacterial cultures. Only the Italian works of Rivera and Sempio are cited here.*

We are no longer moving, as the reader will see, in the sphere of the substantial nourishment of plants, but in a sphere that, for simplicity, we refer to as *dynamic*. Observation of these dynamic relationships of plant life, as well as the substantial, belongs to the biological part of this book. The choice of these highly diluted substances from the environment constitutes dynamic activity on the part of the plant.

The finest example of this activity is the *Tillandsia usneoides* (Spanish moss), a plant growing extensively in the U.S. South from Texas to Florida, as well as in Central America. It is no saprophyte but lives on trees and, more remarkable, can be found occasionally on electrical wires. Its mode of obtaining nourishment is something to make us pause and think.

> This most remarkable of all the epiphytes that, in tropical and subtropical America, often completely unveils the trees, consists of thread-like shoots, often more than a meter long, with slender, grass-like leaves that—in the first stages of the plant's growth—are connected with the bark only by means of early dried, weak roots. The fact that they remain fastened is because of the fact that the basal part of the axillaries winds around the stem portions. The sprouts are covered with scale-like hairs, similar in form and arrangement to those of other *Bromeliaceae*. The spread of the plant occurs less through seeds than vegetatively; sprouts are torn off and carried away by the wind or by birds that like to use them as material for nest building."**

---

* V. Rivera, *Sulla influenza biologica della radiazione penetrante*, Bologna Licinio Capelli-Editore, 1935. V. Rivera, *Ancora sull azione biologica dei metalli a distanza*, Roma Scuola Tipografica, 1933. Caesare Sempio, *Rapporto fra effetti prodotti dai metalli posti a distanza, a contatto e in solfitazione sullo sviluppo della Thielaviopsis basicola*, Pavia, Premiata Tipografia, 1935.

** Quoted from the plant geographer Andreas Franz Wilhelm Schimper.

The most interesting thing about this plant is the way it obtains its nourishment. Since it has no roots, it cannot obtain nutritive substance from the surface on which it grows. Its intake of nourishment is accomplished by means of leaf and stem organs. Wherry and Buchanan describe this process:

> The Spanish moss is an epiphyte, requiring support by other plants and usually hanging from trees. It is not, as sometimes supposed, a parasite, since it draws no nutriment from the sap of its support, actually growing even better on dead trees than on live ones and, moreover, thriving on electrical wires. The scales with which it is covered have often been interpreted as serving to retard transpiration, but Pessin has recently shown that the similar scales of the resurrection fern do not function in this way, and suggests that such scales on air plants more likely serve to hold water by capillary activity while the plant is absorbing therefrom the mineral constituents it requires. The mineral nutriment of the Spanish moss is evidently obtained, then, from whatever salts happen to be present in the rain that falls on it, in water that drips on it from nearby trees, or in dust blown in by the wind. The composition of its ash should accordingly be of interest as affording an indication of the materials added by wind and rain to the soils of the regions in which it grows.*

The two authors then present mineral analyses of the *Tillandsia*, followed by an analysis of the rainwater. In these analyses, two points are especially noteworthy: 1) the high percentage of iron in the plants (averaging 17%), silicic acid (averaging 36%), and phosphoric acid (averaging 1.85%); and 2) The low percentage of iron, silicic acid (H_4SiO_4), and phosphoric acid (H_3PO_4) in the rainwater of the regions where these plants grow—Fe, 1.65%; H_4SiO_2, 0.01%; and no phosphoric acid at all. The authors therefore state:

---

*   E. T. Wherry and R. Buchanan, *Composition of the Ash of Spanish Moss*, Bureau of Chemistry, U.S. Department of Agriculture, from *Ecology*, vol, 7, no. 3, July 1926.

It is evident that the Spanish moss and other air plants exhibit selective absorption and accumulation of individual constituents to a marked degree.... Consideration of the composition of rainwater shows that, in general, this plant does not take up constituents in the proportions present in the water, but exerts a marked selective action.

In a later work, Wherry and Capen go further into the question of the mineral content of *Tillandsia usneoides*:

It seemed desirable to make further analyses to discover whether this plant would show any marked differences in mineral composition when growing on trees and when growing on nearby electrical wires, and to compare it in this respect with other species of air plants growing in the same regions.*

The authors follow with a series of analyses of different *Tillandsia* plants grown under various conditions. We cite no. 6 as a typical example, it being a *Tillandsia* grown on a cypress tree, at Kissimmee, Florida, and no. 7, a *Tillandsia* grown on an electrical wire in the same place. The following table represents analyses of the ash:

No.	Ash	Na_2O	K_2O	MgO	CaO	Fe_2O_3	SiO_2	P_2O_5	SO_3	Cl
6	4.56	15.85	5.81	14.06	12.09	15.30	20.52	2.30	9.28	10.52
7	5.15	12.96	7.75	8.67	13.28	18.60	28.76	2.90	3.27	4.87

Comparison of the two new analyses shows that the plant growing on wire is notably higher than that growing on the trees in ferric oxide and silica, but lower in soda, magnesia, sulfur, and chlorine. The significance of these differences will be taken up after the results of analyses of another air plant are presented.

There are also plants that have the capacity to grow on electrical wires and nourish themselves. Small ball moss (*Tillandsia recurvata*) belongs to these. The authors also give a series of analyses of

---

* Wherry and Capen, "Mineral Constituents of Spanish Moss and Ball Moss," ibid., *Ecology*, vol. 9, no. 4, Oct. 1928.

these, of which we present two. These plants grew in the same place as the others, no: 2 on elm and no. 3 on an electrical wire.

Analyses of the ash of ball moss:

No.	Ash	Na_2O	K_2O	MgO	CaO	Fe_2O_3	SiO_2	P_2O_5	SO_3	Cl
2	5.52	10.72	10.90	6.97	17.19	18.74	25.07	3.78	7.20	4.99
3	4.27	11.19	5.55							

No. 2. Plant from a tree, Kissimmee, Florida.
No. 3. Plant from electrical wires, same locality.

Here, too, there are clear differences, although partly in another direction from those of the previously cited *Tillandsia usneoides*. From still other analyses, it becomes clear that in a dry climate the farther inland we go, the higher is the content of silicic acid. It is universally observed that the content of mineral substances is, in general, a very fluctuating one. The two plants tested had grown under identical conditions, except for their bearer, and were touched by the same rain, dust, etc. The authors of the report conclude:

> The fact that, in the latter case, the rain falling on the plants could have had no previous contact with any source of mineral matter, other than particles of spray or dust carried high in the atmosphere, *brings out in a striking way the ability of air plants to extract relatively large amounts of inorganic constituents from extremely dilute solutions.*

This quotation directs our attention in a positive way to a sphere of plant physiology that has heretofore been observed, but that has so far called forth but little research: the absorption from the air of highly diluted substances of which only very slight traces exist. Under the conditions prevalent in Florida, the Spanish moss, with its rich mineral content, would be an extremely valuable help if used in making compost for fertilizing the soil.

Some years ago, when the author called attention to such facts, he drew the criticisms of a number of scientists to himself. His laboratory research, presented at that time—which, let it be said, had been checked and controlled from many angles—was regarded with amusement. In view of the biases of certain groups toward this problem, there was nothing to be done but call attention to certain phenomena evident in nature. The biases aroused by observations and research of this sort are even more difficult to understand in light of a whole series of examples of this "selective absorption of the highest dilutions," already cited. Nonetheless, it has become "common sense" to speak of the significance and value of traces of iron, copper, zinc, and manganese for plant health. In fact, the circumstances cannot always to be seen so clearly as in the case of air plants, as we usually deal with plants that grow in ordinary soil.

Apparently, farmers do not influence these finer processes directly—yet the health and the intensity of growth depends on them, just as it does on soil cultivation and fertilizing. First, a state of equilibrium arises in nature. This is disturbed by people or by outer circumstances through, for instance, intensive fertilizing with imbalanced minerals that are not "buffered."* However, nature acts to help herself. In the selective capacities of plants, nature herself creates one-sidedness or specialization. But such conditions of one-sidedness arise only when it is their function to heal the effects of another imbalance in nature.

We see no mere chance in this, but wise foresight on the part of nature. We must distinguish in this the bulky, heavily fed plants that grow only in good soil with good intensive fertilizing; these constitute the greater part of our cultivated plants. They collapse in the face of any phenomenon of deficiency. These plants must be distinguished from the dynamically active plants that, on their part, help to correct the unbalanced state of nature. We must study these plants exactly in their relationships to nature, and thus discover one of the important secrets of biological phenomena. We present several examples to illustrate this principle.

---

*   *Buffered* = harmonized through the organic colloidal condition of the soil.

Tobacco is rich in potassium when it grows in a soil poor in potassium and vice versa. The wood and bark of oak trees are especially rich in calcium (up to 60% or more of the ash is CaO). Moreover, they can grow in sand (i.e., in calcium-poor soil), and despite this are able to accumulate calcium. Their iron content might reach 60%. Buckwheat is a typical sand–silica plant and distinguishes itself by its accumulation of calcium. Here we see dynamic–selective conditions.

The broom shrub (*Sarothammus vulgare*) is especially remarkable. It is conspicuously rich in calcium in its stamens and leaves (25.03% CaO where the soil has only 0.35% CaO). Moreover, it gives off calcium from its roots, depositing it in encircling rings, thus providing the soil with calcium. This phenomenon—occurring in a predominantly sand plant—garners special attention. These qualities make this plant especially suited to helping to prepare dry wastelands and sandy prairies and, by enriching their calcium content, bringing them back to a state of cultivation.

Daisies might appear on lovely English lawns, which constantly increase in acidity and decrease in calcium. Daisies are remarkably rich in calcium and, at the same time, indicate that the soil has passed a certain "acid limit"; it is their task to collect the substance it needs. In such cases we might ask: *What is the source of this substance? How does the plant take it up?* So-called weeds become enormously significant in this connection. They are "weeds" only from a human, utilitarian point of view. At most, we might call them "misplaced" good plants, or good plants growing in the wrong place. A more exact examination, however, shows that they are very much in the right place. They are the most accurate specialists; they have adjusted to a specific degree of acidity, so that their mere presence gives us a precise indication of the soil's acidic or alkaline condition. "Weeds" are the traffic lights and warning signals of soil life. Let's cite a few more of their specific qualities, so that we might reach a conclusion from all this.

Relationship to magnesium: Fagus silvatica becomes rich in magnesium with increasing age, with 12.4% in a 10-year-old, up to 19.5% in a 220-year-old tree (shown in the ash). Oak and *Picea excelsa,* on the contrary, show a decreasing magnesium content with increasing age (the oak, for example, shows 13.4% at 15 years and 2.35% at 345 years). Barks of trees are in general poor in magnesium; birch, on the contrary, is conspicuously rich in it, showing often as much as 14%. Normally its leaf ash contains from three to eight percent.

Specialists in accumulating magnesium are:

	% of MgO content		% of MgO content
Prunus	12.3	Solanum tuberosum	28.5
Acer campestris	10.5	Stellaria media	21.8
Erica carnea	15.5	Ilex aquifolium	20.6
Betula alba	15.3	Herniaria glabra	18.9
Scrophularia nodosa	15.6	Spiraea ulmaria	18.0
Beta vulgaris	25.9		
Fat-accumulating seed pods in general contain more MgO			
Grain varieties	11.0	Linum	14.1
Maize (corn)	15.5	Gossypium	16.6
Fagopyrum	11.4	Cannabis	5.7
Pinus	7.9	Theobroma	11.0
Phaseolus	7.6	Cocos	9.4
Castanea.	7.47	Aleurites	15.1
Quercus	5.2	Juglans	13.0
Brassica rapa	11.8	Abies	16.7
Brassica napus (rape)	13.4	Amygdalus	17.7
Papaver	9.4	Wood ash in general	5–10

Worthy of special note are: *Rubus fruticosus,* 15.81%; *Betula,* up to 18%; *Quercus,* 15% to 23%; *Larix,* up to 24.5%. In all these, there are also rhythmical variations in the course of the year.

Farmers might ask: *What significance does all this have for me?* The significance is that many of these plants live in or near their cultivated areas, and they sometimes fertilize the soil by dying off, by dropping their leaves, and so on. More important, they furnish, in available organic form, exactly the finely diluted substance that nature needs for healing or stimulating its life processes. The remarkable thing is that many of these plants are "specialists" in producing the substances the soil lacks, thus improving the soil. As we've seen, they accumulate these substances from states of high dilution and, by condensing them into more concentrated form, subsequently carry them into the soil.

A synthesis of this process can be accomplished in a practical way by making compost of everything. The greater the variety of plants compost, the richer and more useful its nutritive potential. Compost made of weeds or medicinal herbs (in this case, they are closely related) can carry into the soil what cannot be brought in another way that is organically in harmony with it. Since we are dealing largely with dynamic effects, the most minute quantities of one substance or another can often be sufficient.

It might be interesting and valuable to mention a few more special plant activities. For instance, henbane (*Hyoscyamus niger*) and jimsonweed, or thorn apple (*Datura stramonium*), like to grow in calcium-rich soils. Their ash content is relatively poor in calcium, whereas they both "specialize" in forming phosphoric acid—henbane with 44.7%, and jimsonweed with 34.7%. Common foxglove (*Digitalis purpurea*) likes iron, as well as calcium and silicic acid; it also stores manganese.

Dr. Fahrenkamp observed that substances from red foxglove, lily of the valley, and plants of the scilla genus that affect the heart have a preserving influence on other plants.[*] Freshly cut flowers placed in water that contain traces of these substances, or a bouquet of flowers containing a few of these plants, keep much

---

[*]   Karl Fahrenkamp, *Hippokrates,* Stuttgart, 1937.

longer than usual. They even advance the flowering and maturing processes of plants.

Thus, again we might ask whether or not plants such as weeds have an important influence on their surroundings through their secretions, even when highly diluted. If, in addition to this, the research of Hans Molisch is considered—according to which the "exhalation" of a few apples (the result of a presumed trace of ethyl) causes plants growing nearby to mature more quickly and shed their leaves earlier, while the same "exhalation" checks germination and root growth—we can see a new field of research open up that has, thus far, been neglected and even ridiculed.* Molisch's research was carried out in an especially large number of cases and include a study of the ecological influence of plants such as apples, pears, peas, vetch, broad beans, potatoes, and, in addition to these, tangerines, oranges, lemons, bananas, cherries, apricots, green gage plums, *Ceratonia,* carrots, beets, Horseradish, and many others.

Dandelions are fond of calcium and silica, whose ash residue is especially large in these. It grows luxuriously in countries where alfalfa is cultivated. *Robinia pseudacacia,* or black locust (the significance of which as a basis for afforestation we treated in chapter 8), prefers sandy soil and accumulates calcium up to 75%. Its leaves in a mixed forest are the best fertilizing agent against calcium deficiency. Moreover, it provides nitrogen to the soil.

Another interesting plant is the sea beet (*Beta vulgaris maritima*), which grows near the sea. It is a mini-pharmacy, with a composition of up to 56% sodium, lithium, manganese, titanium, vanadium, strontium, cesium, copper, and rubidium. Where fertilizing is imbalanced, the multifaceted nature of the sugar beet disappears. How could a sugar beet be comfortable under these conditions? Sea beets grow along seashores, deeply indented bays,

---

*   Hans Molisch, *Der Einfluss einer Pflanze auf die andere: Allelopathie.* Verlag Dr. Müller, 2007.

and fiords, where they are fertilized by nature with a compost of seaweed, which is in turn a pharmacy itself that carries many of the ingredients the sea beet likes. Wouldn't we make sugar beets happy by providing a few small doses of compost containing seaweed or its ash, served up as a dressing, to remind it of its primeval origin? Boron, for example, is also present in such seaweed compost. Thus, we would help the sugar beet far more, and in a more natural way than by any artificial treatment that lacks the proper feed elements.

The corn marigold (*Glebionis segetum*) is a weed that prefers calcium-deficient loamy and clay soils. It is rich in calcium, phosphoric acid, and magnesium. This should be turned to advantage.

Red sorrel (*Rumex acetosella*) can be found on sour parcels of land. It warns us when pasture soil is becoming too matted and closed up. Its ash is rich in calcium, phosphoric acid, magnesium, and silicic acid. *Into the compost pile with it!* Let it fertilize the same pasture again—after harrowing.

Horseradish (*Cochlearia armoracia*) is a supplier of calcium (11.9% Ca), phosphoric acid (13% P_2O_5), and sulfur (18.64% So_2). Potatoes also do well in its area.

Chamomile (*Matricaria chamomilla*) is generally rich in salts, especially potassium (45%) and calcium (23%). It is conspicuous is its richness in sulfur compounds.

Cacti contain much calcium, and certain varieties accumulate lime up to 80% of their dry substance.

Horsetail (*Equisetum*) varieties are strong gatherers of silicic acid, like grasses that have leaves rough to the touch when stroked downward from tip to base.

Yarrow (*Achillea millefolium*) is rich in potassium, calcium, and silicic acid.

Stinging nettle (*Urtica dioica*) is rich in calcium (36.4%) and silicic acid.

Annual (or burning) nettle (*Urtica urens*) shows incrustations of silicic acid on its stinging prickles.

The onion is a specialist in silicic acid, with calcium in its leaves.

These examples represent only a small sample of this field of study, much of which remains untouched by research. It is essential to direct biological study toward this area of practical importance. Nonetheless, we are certain of one thing—the more varied the compost, the more certain its dynamic effects. Some of these plants are also important in directing the fermentation of organic matter.

In chapters 7 and 8, we have already treated the significance of a mixed culture. The presence of weeds must be considered as a dynamic factor. Garden orach (*Atriplex hortensis*), or mountain spinach, prefers to grow on humus soil treated with good organic fertilizer. This plant has an especial affinity for the potato. While the potato otherwise chokes out root herbs, orach likes to grow next to it and checks its growth. It can even happen that soils exhausted by potatoes indicate this fact by showing an increase in the growth of orach, just as soils exhausted by hoed crops in general frequently show this by an increase of black nightshade (*Solanum nigrum*); tired soils that become crusty because of constant planting of grains show it with an increase of chamomile (*Matricaria chamomilla*), and soils overfertilized with potassium show it by the spread of hedge mustard (*Sisymbrium officinale*).

The mutual influence of plants belongs, in any case, to the borderline spheres of dynamic and biological effects. When wheat and poppies are grown together, the poppies check the development of the wheat and thus lower the wheat kernel yield, whereas corn flowers are less harmful.

This author has conducted experiments for years in connection with plant symbioses. It was found that chamomile, for example, can be both beneficial and harmful. When there were only a few chamomile plants growing among or along the border of an experimental field planted with grain, the chamomile actually fostered its growth. If, however, it is planted thickly, for example in rows among the grain or along the border, then it has an inhibiting effect. Here we are dealing with a typical dynamic effect.*

In further experiments, it was possible to observe other stimulating and inhibiting effects. Some of the tests were arranged as plant experiments in an open field, others as seed baths with extracts of the relevant plants (a half-hour bath and then sowed and grown normally), some as germination tests after the seed bath, and yet others as tests of growth in a seed bath of corresponding dilution. In practice, we can use our knowledge of the effects of so-called border plants, sown along the edge of a bed or field of cultivated plants. Indications and the impulse for this investigation came from Rudolf Steiner and have proved to be of great value.

Tested by many different methods, the following have shown themselves to be especially helpful—dead-nettles (*Lamium*) and the sainfoins, or esparsette (*Onobrychis*). Another good border plant is valerian, and here and there yarrow. Harmful examples are too much chamomile, buckwheat, hemp, and cornflowers, as well as the poppy.

Lippert's research shows the significance of the stinging nettle. Planted in rows between medicinal herbs and compared to the control plots, it raises the content of etheric oils in those plants. It further proves that the pressed juices of plants grown alongside stinging nettle are more resistant to mold and putrefaction.

From other literature on the subject, the following is also familiar.** Rye is hostile to weed growth, inhibiting germination and growth of field poppies among other plants. Where there is a strong

---

* Cf. Friedrich Boas, *Über Hefewuchsstoffe:* chamomile, in a dilution of 1:8,000,000 stimulates yeast growth. Other substances stimulating yeast growth are effective in dilutions of 1:4,000,000.

** Dr. J. Kubn, *Mitteilungen der Deutschen Landwirtschaftsgesellschaft,* 1932, vol. 1.

growth of couch (twitch) grass (*Elymus repens*), rye planted twice in succession can eliminate this troublesome weed. Poppy and larkspur (*Consolida*) show an affinity for winter wheat but not for barley. The seeds of these weeds lying in the ground awake to activity when, in the rotation of crops, it's winter wheat's turn.

Hedge mustard (*Sisymbrium officinale*) and field mustard (*Sinapis arvensis*) are fond of oat fields, and in this case they signify no inhibiting symbiosis. On the contrary, they both work to inhibit in the case of rape. Especially harmful for turnips are the hedge mustard and the knotweed (*Polygonaceae* family). Red clover (*Trifolium pratense*) and plantain (*Plantago*), alfalfa and dandelion like to grow together. The effect of rye on wild pansy (*Viola tricolor*) is extraordinary. While the latter in general only germinate from 20% to 30%, they germinate up to 100% in a rye field.

If these facts are applied in the right way, planting border and protective plants can be developed and greatly assist in combating the effects of present-day, one-sided field cultivation and become especially important in cultivating gardens and growing medicinal plants.

The purpose of this chapter is to illustrate what may be understood by "the biodynamic principle in nature" and to show how we can learn to make use of it.

CHAPTER 11

Scientific Tests

There are two arguments by opponents against the biodynamic method of agriculture. One has to do with the effect of the preparations. Special handling of manure and compost (or so we hear and read) is described as exemplary, and recently it has even been referred to as self-evident. But considerable doubt is expressed regarding the effectiveness of those preparations or whether they have any special effect. The other objection has to do with the long-time effect of the biodynamic method. If we were to work in that way we would be "robbing" the soil of its nutritive materials, so the objectors say. We have already discussed the second question in chapter 2. Chapter 12 provides practical agricultural results to show this fear is unfounded.

A series of experiments has been carried out by the author and his associates in the chemical–biological research laboratory at the Goetheanum. The simplest experimental procedure is the following: Seeds, chopped potato, and such are bathed for a half-hour to an hour in a thin dilution of preparations 500, 501, 502, 503, 504, 505, 506, and 507 (all made according to directions by Rudolf Steiner). Above all, the development of the roots is observed. A variation of this method is to let the plants grow directly in the corresponding dilute solution.

In every experiment performed this way, results show that preparation 500 has an especially stimulative effect on the root growth, causing the development of numerous fibrous roots. Preparation 501 increases the assimilative activity of the plant. Preparations 502 and 507 strengthen plant growth in general. In this connection, 504 has a special influence on the quality of flavor, the others more on the production of mass. The exact results of these experiments were published in the monthly periodical *Demeter,* published in Bad Saarow, near Berlin.*

We present here two experiments illustrative of these results. Lupine was germinated in a water culture with the addition of the various solutions and planted in soil. This was followed by irrigating all the plants with the same tap water. After 10 days, the length and weight of the plants were measured. We had, as before, control plants germinated in tap water alone, other plants germinated in baths of preparation 500, and others with 501 in a 0.005% dilution, and another group of plants germinated in the 500 solution, their leaves sprayed with 501 (seven days after planting) in the dilution used in ordinary practice (given in the table as preparation 500 along with 501).

Mean values for 20 plants in each case	Control	500	500–501	501
Root length after 19 days, in cm.	11.08	13.74	15.16	14.5
Root weight after 19 days, in gr.	0.363	0.443	0.726	0.452
Sprout length after 19 days, in cm.	9.5	10.9	13.0	12.5
Sprout weight after 19 days, in gr.	2.071	2.093	2.426	2.42

Result: Plants treated with the preparations are superior to the control group. The arrangement used in general practice: preparation 500 for germination, with preparation 501 sprayed on the leaves after transplanting; results, maximum.

An experiment with wheat: Germination and growth in sand, containing humus. Growth measured after 12 days. The soil of the control was moistened with tap water, the others had respectively

---

* Pfeiffer, *Pflanzenversuche zur biologisch-dynamischen Wirtschaftsweise,* no. 7, July, 1931; Pfeiffer, Kunzel, Sabarth, *Versuche zur Wirkung der Präparate 500, 501, sowie 502–507,* no. 7, July, 1935.

500, 501, and 500 plus 501, in the previously given dilutions. Preparation 500 was applied before sowing, preparation 501 was sprayed on the leaves.

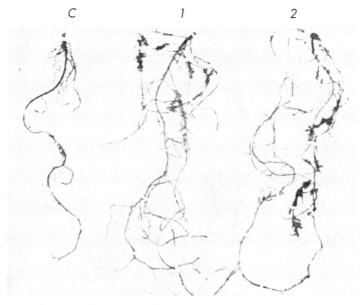

C—control-germinated in tap water
1—pre-treatment with 500
2—pre-treatment with 500 and again before transplanting

Mean values for 30 plants in each case	Control	500	500–501	501
Sprout length in cm, first leaf	11.8	10.7	10.5	11.6
Sprout length in cm, second leaf	5.6	4.7	6.1	6.5
Root length in cm	17.2	22.7	20.2	22.0

The result corresponds to previous experience.

Let's add pictures of typical test plants to the figures on the experiments with white lupin (*Lupinus albus*). In these, the forms of the roots are especially noteworthy.

3   4

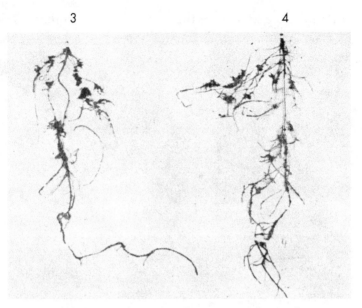

3—*germinated in 500, 501 sprayed on leaves (note in 2–4 the strong
    development of the root hairs and the bacteria nodules)*
4—*Germinated in 500, 500 at transplanting, later 501 sprayed on leaves*

5   6   7

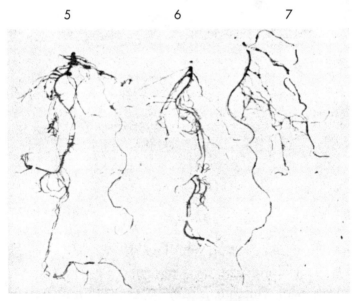

5—*Germinated in 501, long, thin root*
6—*Germinated in 501, 501 again at transplanting*
7—*Germinated in 501, 501 again at transplanting,
    501 sprayed on leaves (wrong treatment in cases 6 and 7)*

Incorrect, premature use of preparation 501, especially in number 6, leads to atrophy and weak root development.

Special attention was given to the study of the stimulative effects of preparation 500 used together with light applications of biodynamically treated compost.

The value of nitrogen bacteria for the legumes and for general fertility of the soil is well known. Hence, it was important to determine whether the formation of nodules on the roots of the legumes is influenced by this way of working. For this experiment we used field soil that had been fertilized biodynamically for a number of years and other soil from the same kind of ground, which had, however, been treated for a number of years with so-called complete fertilizing (organic mineral fertilizing). Equal amounts of these two soils were separately mixed with pure sand. The soil was a heavy clay and had to be made somewhat looser by an addition of sand.

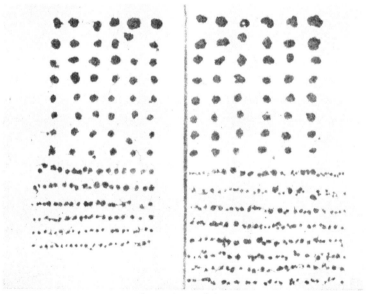

Bacteria nodules

Then the mixtures were put into flowerpots and at the same time a planting of common pole beans (*Phaseolus vulgaris*) was sowed in both. The seed was carefully selected and equal in weight. The experiment, carried out in the greenhouse, was concluded after the beginning of the fruit formation. The plants were carefully freed of the soil, washed, and the bacteria nodules carefully separated from the roots. The nodules of from 2 to 4 millimeters in size were counted, and then the total weight of all the nodules (those smaller than 2 millimeters, as well as the rest) was determined.

Result: weight of the root nodules in grams, calculated on the basis of 100 plants:

Biodynamic	*16.2 gr.*
So-called complete fertilizing	*9.5 gr.*

It is clear from the table that the nodule formation is substantially benefited through the biodynamic fertilizing. If we consider the high economic value of the legumes in agriculture, especially their function in the crop rotation, a wide perspective opens up from this experiment. The fixation of the atmospheric nitrogen (which cannot be taken up directly by the plants) into organic nitrogen compounds is accomplished by the legume root nodules. Furthering this activity means enriching the soil in nitrogen compounds that can be taken up by the plants. We see here one of the primary merits of the biodynamic method of agriculture—that it furthers the natural activity of soil life in a way that has not been observed when other fertilizers are used.

The significance of earthworms for humus formation in the soil has already been discussed (chapter 4). It has come to light in practice that the biodynamic preparations especially improve the general conditions in the environment of earthworms. Hence, the earthworms migrate toward these more favorable conditions. This can be proved by simple laboratory tests.*

We now come to a series of experiments that took place in the research laboratory at Dornach. The procedure was as follows: A wooden box was divided into four equal compartments filled

---

* J. v. Grone-Gueltzow, *Wie verhält sich der Regenwurm zu biologisch gedüngten Boden. Gaia–Sophia.* vol. 4, Dornach, 1929.

respectively with the same earth treated in various ways. One portion had been watered with a customary dilution of fertilizers containing potassium, nitrogen, and phosphoric acid. The second was treated with a dilution of urine. In the third division, soil was deposited that had been fertilized with our biological preparations. In the fourth, the same soil was unfertilized as a control. Of course, the same compost soil was the basis for all four tests, and all four were kept equally moist during the experiment. Into each part we placed the same number of earthworms, and after an interval of several days, a checkup was made to determine the location of the worms.

There were small openings between the individual divisions, so that the worms could pass through them from the soil in which they were originally placed, so that (as we have the right to assume) they could choose the soil that suited them best.

Worms set out		Found after 1 day	Increase or decrease
Artificial fertilizer	11	9	−2 = −18%
Urine	11	8	−3 = −27%
Biodynamic	11	13	+2 = +18%
Control	11	12	+1 = +9%
		losses: 2	

Worms set out		Found after 3 days	Increase or decrease
Artificial fertilizer	10	5	−5 = −50%
Urine	10	5	−5 = −50%
Biodynamic	10	24	+14 = +140%
Control	10	5	−5 = −50%
		losses: 1	

Worms set out		Found after 4 days	Increase or decrease
Artificial fertilizer	10	8	−2 = −20%
Urine	10	2	−8 = −80%
Biodynamic	10	22	+12 = +120%
Control	10	6	−4 = −40%
		losses: 2	

We see that the result favored biodynamically fertilized soil. The urine and the artificial fertilizers were readily dismissed and the biodynamically fertilized soil was much preferred by these creatures that, boring their way through the ground, taste with their soft, slimy skin. Beyond the simple numbers, which certainly speak in plain language, we were also able to see how the "biodynamics" quickly disappeared into their soil, especially in contrast to the worms in the urine-saturated soil, which remained lying inert for quite a while on the surface.

It is especially noteworthy that the soil itself in which twenty-four rain worms gathered had a clearly different and greatly improved structure after three days. It is also worth noting that preparation 507, a specially prepared extract of valerian, is attractive to earthworms—not just cats! As we know, normal biodynamic practice is to finely spray this preparation on completed compost or manure piles.

In a test of the effect of 507 on earthworms, the following results were obtained after 5 to 8 days. In four communicating compartments, half were filled with earth, others with earth plus 507. We can gather from this that the spraying of the heaps with preparation 507 has an effect of attracting the microorganisms, and so on of the soil, and brings them in swiftly to start their activity.

	Worms at the start of the experiment	Worms at the end of the experiment
Experiment one:		
Earth	10	9
Earth plus 507	10	18
Experiment two:		
Earth	10	8
Earth plus 507	10	19*

*Additional worms moved in from outside

Spade-full of three-month-old biodynamic compost with earthworms

*Root growth: chrysanthemums that have grown down into the biodynamically
fertilized earth, with the main mass of their roots on the right side*

Another experiment was performed according to the classic method of Dr. A. Wigmann and L. Polstorff, as mentioned in "The Inorganic Constituents of the Plant" (1840). This, it might be remembered, was honored by the University of Göttingen in 1842. In its time this method served for a determination of the nourishment and the mineral salt requirements of plants, out of which then came consequences of such far-reaching importance for mineral fertilizing. The method rests on giving a plant the possibility of letting its roots grow simultaneously in variously fertilized soil and then determining in which direction the preferred growth goes.

nothing added | preparations added

Diagram showing the method of the experiment

Applied to our problem, flower pots were filled with ordinary earth. The bottoms of the pots were removed and placed on the edge of partitions that divided the box into two compartments. Thus, they occupied equal spaces over each compartment. Placed in one compartment was earth fertilized with ordinary compost; in the other, the same earth fertilized with the same compost was placed, but the latter had been prepared biodynamically. When plants were set out in the pots and permitted to grow, the roots grew into one or the other half of the box (see image).

Thus, the plant itself demonstrated an ability to find the conditions of life most satisfactory to it. For practical purposes, the final and surest criterion when investigating a method of agriculture must always be indicated by the way the plants themselves react.

The experimental arrangement just described was also used to show the individual effects of the preparations. Test plots were divided in two sections by a separating glass partition. The partitions were covered with untreated compost; but, at a lower level, on one side of the partition biodynamic preparations were inserted, on the other side of the partition no preparation was used. The plants grew first in neutral earth, then they continued their growth, their roots working down into the variously treated soil, and developing variously in accordance with this. With this experimental arrangement we were able also to confirm the effect of the preparations.

Rate of growth of soybean roots in biodynamically treated earth and in the same soil untreated

Preparation #	Average % of increase compared to control	
	Length	Weight
502	+ 4.5	+ 1.6
503	+ 12.8	+ 6.9
504	– 5.2	– 13.8
505	+ 4.5	+ 14.2
506	+ 4.5	+ 22.7
507	+ 5.1	+ 10.8
502–507	+ 5.8	+ 20.5

This table shows a whole series of interesting results and plainly demonstrates one thing—that the preparations definitely produce effects. In one case—the 504—notice that the quantitative effects are inferior to the control. However, it must be noted in this connection that other researchers have proved that this preparation causes an intensification of qualitative characteristics such as aroma and the like. The table also shows that the effects on root growth and weight are quite varied. It is clear that here we are not dealing with an ordinary growth stimulus, but that the whole structure of the root—its density and water content—were also altered.

Seed-bath tests with biodynamic preparations show similar differences. The seeds used in the experiment were placed for 20, 30, or 60 minutes in a highly diluted solution (e.g., 0.005%) of each preparation, dried, and then immediately planted. It was found that, when the seeds were permitted to lie for several days after the bath, the beneficial effect was lost. Such seed-bath experiments have been made for several years in the author's laboratory and elsewhere and are sufficiently extensive to allow a definitive judgment of the preparations' effects. Since only small quantities of material in high dilution are used in these tests, and for only brief periods of time— after which the seeds are planted in the same, ordinary soil and grown under normal temperature and moisture conditions—*here it is clearly a matter of dealing with a purely dynamic effect.*

Test with radishes: average value for 20 plants in each case. The experiment is closed when the radishes are ripe: (a) weight of radishes including roots, (b) weight of leaf mass.

	(a) in grams	(b) in grams
Control	2,570	2,530
Seed bath with preparation 500	2,775	2,655
Seed bath with dilution of prepared manure	3,010	2,300
Seed bath with preparation 500 + 502–507	4,600	4,090

Such seed-bath experiments show very markedly the effect of the preparations in stimulating growth. Here matters must be confined to a presentation of principles. The complete results of the experiments in this field will be published at some future time.

Test with sweet corn: average value of 10 plants in each case

	Height of plant in cm.	Weight of roots in gr.	Leaves and stems in gr.	Weight of ears in gr.
Control	181	50	2,181	727.5
Seed bath with 500 + 502–507	185	65	2,250	800

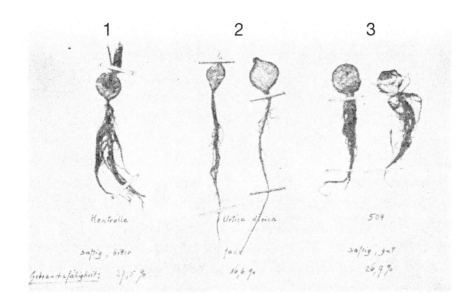

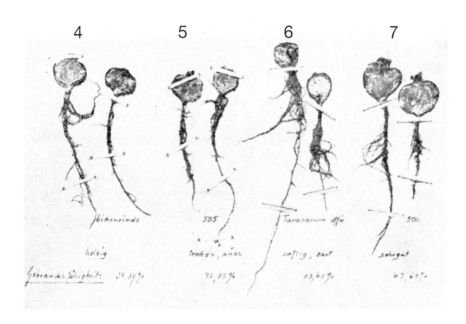

Roots of radishes with the preparations
and roots of the controls are typically different

It must especially be stressed that the plants used to make the biodynamic preparations would not produce the same effect on their own before being put through the biodynamic process of preparation that they produce after this treatment. Comparisons have been made of seed baths containing extracts of the selected fresh plants with seed baths containing extracts of the same plant materials after they have been put through the biodynamic process. The variations of growth and root development on the tested plants have been carefully observed and noted. In addition, a new factor of "usability" has been set up—a compound of flavor and uniformity, size and shape. From the results, we see first that the effects of the preparations are very varied, and second, that this variation shows itself in the quantitative and qualitative aspects of the plant treated with them. *It is important to note the fact that the very brief seed-bath treatment influenced the later growth of the plant.*

The forms of the tested radishes and control plants show very characteristic differences (see images on page 145).

1. Control:	Taste – juicy, bitter	Usability: 27.50%
2. *Urtica dioica* (stinging nettle)	Taste – insipid	Usability: 16.6%
3. Preparation 504 made of urtica	Taste – juicy, stimulating	Usability: 26.9%
4. Oak bark	Taste – woody	Usability: 34.37%
5. Preparation 506 made from oak bark	Taste – dry, sweet	Usability: 42.85%
6. *Taraxacum* (dandelion)	Taste – juicy, delicate	Usability: 13.63%
7. Preparation 506 made of dandelion	Taste – stimulating, aromatic	Usability: 63.63%

Another result of the experiment: It showed that the yield in mass and the yield in quality do not always coincide.

Experiments with the biodynamic preparations, carried out by the Greek Government Office for the Protection of Tobacco in Thessaloniki, show interesting results both quantitatively and qualitatively as compared with artificial fertilizers and untreated soil. The preparations—502, 503, 505, and 507—instead of being used in compost and manure in the usual way were stirred into water and sprayed directly into the soil (as directed for preparation 500). Preparations 500 and 501 were employed in the usual way. The tobacco yield, with no other manure treatment, was superior to that obtained with a dressing of Ammophos (14–28–14),* and only slightly inferior to the yield of a nitrogen–phosphoric, acid–potash fertilizer (6–8–8). Growth was rapid and regular, and leaves were erect and deeply colored.

Experiment 1

	Manurial treatment	Yield in "okes" (per 500^2 meters)
Plot # 5	Nitrogen, phosphate, potash (6–8–8)	47½
Plot # 3	Preps. 500, 501, 503, 505	45
Plot # 6	Prep. 500	44
Plot # 1	Ammonium phosphate (14–28–12)	42
Plot # 2	Control (no fertilizer)	35
Plot # 4	Control (no fertilizer)	33

Experiment 2
Another experiment with biodynamic preparations yielded these results

	Manurial treatment	Yield in "okes" (per $2,500^2$ meters)
Plot # 3	Preps. 500, 501, 507	190
Plot # 1	Prep. 500	132
Plot # 2	Preps. 500, 503, 505 (hail damaged)	122
Plot # 4	Control (no fertilizer)	121

A preliminary report on the quality of the tobacco, received before the process of fermentation was completed, gave the following results:

---

*   *Ammophos* is a highly concentration nitrogen–phosphorous fertilizer, produced in various grades and might include additional microelements.

A: Quality of color

Plot # 3	Preps. 500, 501, 507	1st
Plot # 1	Prep. 500	2nd
Plot # 2	Preps. 500, 503, 505	3rd
Plot # 4	Control (no fertilizer)	4th

B: Quality of aroma, texture, etc.

Plot # 3	Preps. 500, 501, 507	1st
Plot # 1	Prep. 500	2nd
Plot # 2	Preps. 500, 503, 505	3rd
Plot # 4	Control (no fertilizer)	4th

Full conclusion of the report

1) The biodynamic preparations have given satisfactory results as regards quality.

2) By the use of the preparations the tobacco plants developed more rapidly and reached maturity some days earlier.

3) Plot no. 3 (500, 502, 507) gave the best combined results. Plot no. 2 (500, 503, 505) came second. This plot—notwithstanding its seemingly small yield—competed effectively with the previous plot in both quality and quantity. We believe that, if this plot had not suffered from severe hail that destroyed 40%, it might have been the best of all. Third place was taken by plot # 1 (prep. 500) while the control plot came in last.

Contrary to our initial opinion, we believe—or better, we are convinced 1) that these preparations, supplemented rationally and utilized with care and attention, can become beneficial in every respect to the cultivation of tobacco; 2) that it is indispensable to continue the experiments with all the preparations for at least 2 more years to enable their inventor and us to draw more precise and positive conclusions for the benefit of the Greek tobacco industry.

In carrying out of such experiments, it must be added that a rich, heavy soil produces less variation in the results of the experiments than when they are done in a poor or light soil. The different seasons of the year have to be considered, because any condition that causes luxuriant growth will lessen the experimental variations.

It is clear from these experiments that a stimulating effect on growth comes from the preparations. Because of this, the small amounts used are sufficient. For example, we sprayed 40 grams of preparation 500 diluted in 10 liters of water on one acre, or we insert one or two grams each of preparations 502 and 507 into one to three cubic yards of manure.

We know from the study of vitamins, hormones, and ferments that increased effects can be produced by small doses of certain biologically active substances. But if we attempt to apply this knowledge in a practical agricultural way, we often encounter profound amazement in those who are supposedly practical; it's as though there could not possibly be progress in this area! Naturally, we have to begin with the proper foundation—i.e., by providing organic fertilizing and treatment of the soil—because it is this soil we want to stimulate to increased productivity by using the knowledge given by these results.

The effect of biodynamic treatments shows itself not only in the changes of manure and soil; it also extends to the quality of the food and fodder grown in it. In the case of wheat, an increase was observed first in its germinating capacity and germinating energy.

The result of six germinating tests with 350 grains of wheat

Kinds of wheat	Wheat germinated in distilled water	Weight increase of seedling after 14 days (measuring only green plants)
Biodynamically grown	99.3%	98%
Wheat (same species) grown in identical soil but according to previous methods (i.e.,with fertilizing materials but no biodynamic treatment)	92.3%	86%

With wheat, another difference also showed itself in gluten content and gluten quality. With Dutch Juliana wheat, grown in neighboring test fields, the following was demonstrated:

Gluten content: Juliana wheat first year before the biodynamic treatment (original seed) 30.6%
  Gluten quality: elastic, well-formed of regular structure and normal rising. energy

Gluten content: wheat second year, after growing with biodynamic treatment 41.0%
  Gluten quality: elastic, well formed, excellent rising capacity

Gluten content: wheat the second year, same kind of soil but with mineral fertilizer. 28.8%
  Gluten quality: medium elasticity, irregular structure, medium rising capacity

Gluten content: Carsten V wheat (original seed) 22.5%
  Gluten quality: well formed, weak rising capacity

Gluten content: Carsten V, 1–2 years biodynamic treatment . 23.6%.
  Gluten quality: elasticity and form capacity increased, rising energy strengthened

CHAPTER 12

Health Effects of Fertilizing

Scientific tests have demonstrated the general improvement in the baking quality of biodynamically grown wheat in contrast to wheat grown under other conditions. This has led to making biodynamic products, such as special flours and breads. In Europe, the products raised and made thus carry the trademark *Demeter*.

Through feeding experiments with animals, it has been possible to observe qualitative differences. The tests were made with white mice. Three strains of mice were fed biodynamic grain, three with mineral-fertilized grain of the same sort. In every generation, two pairs were raised to maturity from each strain. Thus, in each generation, six litters were nourished with mineral-fertilized grain and were available for comparison to six litters that were nourished with biodynamically fertilized grain.

Raising the animals was carried out as follows: At four weeks, the young were separated from the mother and segregated according to sex. After this, the animals were weighed regularly. At nine weeks (when they are typically mature), they were paired so that each new generation was born with great regularity at 12-week periods. The most identical conditions possible for the experimental animals were created by this arrangement.

The food used was Ackermann's Bavarian brown wheat, the seventh sowing of this strain, biodynamically fertilized for five years, and the same wheat mineral-fertilized with kainite and basic slag in the customary amounts. All the original wheat used in these experiments came from the same region and was therefore grown under the same climatic and soil conditions. Feedings were arranged on a sliding scale based on weight. In addition, the animals received a ration of one part milk and two parts water. The milk was boiled for 10 minutes.

The results of the experiments gave the following points: The average number of animals in each litter was:

For those fed with mineral-fertilized grain 6.2

For those fed with biodynamically fertilized grain 6.7

There was thus a slightly larger number of animals born in the case of those fed biodynamically.

Average weight of animals	Minerally fertilized grain	Biodynamically fertilized grain
At four weeks	7.9 gr.	8.5 gr.
At nine weeks	17.2 gr.	16.4 gr
Died at 9 weeks or earlier	16.9%	8.6%

These figures are an average taken from 3 generations, with a total of 164 animals.

Since the number of deaths in the period from birth to maturity can be regarded as a measure of the powers of resistance, these figures show that the animals fed with biodynamically fertilized wheat were notably stronger, although in their weight they were a little inferior to those nourished with mineral-fertilized grain. Regarding the cause of the deaths, it must be added that no sickness arose during the experiments, but that the deaths were due rather to the number of weaklings in the litters that, being hindered in their development, died off.

Following this, to determine the influence of feeding on successive generations, an experiment was undertaken with chickens that continued for a year. The experimental animals (white leghorns) were hatched June 28, 1931. On October 19, they were separated into two identical pens—B and M—with adjoining runs. There were seventeen hens and one cock in each.

Beginning on June 28, pen M was fed daily 100 grams of mineral-fertilized wheat per bird (the Wilhelmina variety). Pen B was fed daily 100 grams of biodynamically fertilized wheat per bird (the same variety, all grown on identical ground). Both pens received the same kind of mash and cracked corn (not biodynamic) to the amount of 50 grams per chicken each day. The only difference in the feeding was the use of differently fertilized wheat.

On February 1, pen M received, in addition, a daily ration of ordinary, ground limestone for feeding, one to two grams per bird. Division B received the same amount of Weleda ground-feeding limestone.*

Pen B began to lay eggs on December 11, and Pen M December 26. The difference between groups M and B in the course of the experiment was as follows: B laid more eggs than M. Individual eggs were lighter on the average in Pen B than in pen M, but the daily total weight was higher for B because of the greater total of eggs.

Average weight of one egg per hen per day:
    M: 59.4 gr.   B: 58.1 gr.
Average total weight of the daily egg production of 10 hens:
    M: 427.2 gr.   B: 464.5 gr.

In pen B, the chickens stayed outside in the run 1 to 2 hours longer in the evening—i.e., their vitality was increased, and with it their liveliness and desire for food. In rainy weather, the birds in pen M sought the shelter of the chicken house sooner than those of pen B.

---

* Weleda ground feeding-limestone is prepared according to a special formula of the veterinary medicine division of the Weleda company.

Total egg production of 10 birds for nine months:

M = 1,495   B = 1,916

Average per chicken in 9 months:

M = 150   B = 192

Weight of the average total daily egg production in grams:

M = 427.2   B = 464.5

A repetition of the experiment in 1933 brought results similar to those of the 1932 tests. The egg production of coops of 10 chickens for seven months in 1933 was: for the B coop, 1,213 eggs, for the M coop 977 eggs. The average egg weight was 64.4 grams for B as against 61.7 for M.

In connection with these tests, certain further experiments were carried out that show that the biodynamically raised feed was beneficial to the animals in every respect. The volume of the egg production was not attained at the cost of the quality of the eggs; on the contrary, the qualitative characteristics corresponded with the improvement in quantity.

Comparative Hatching Test

Forty eggs each of the M and the B chickens were gathered and used for hatching purposes. A "Viktoria" incubator was used for all the eggs, in order to have exactly the same conditions for both groups. The eggs were in it from April 22 to May 12. The eggs had identical treatment.

M eggs hatched	35%
B eggs hatched	68%

Test for the Keeping Quality of the Eggs

Quality is not only expressed in taste. Products of poor quality spoil more quickly because, having less active life forces, they decay more easily. Thus, to test the quality of the M and B eggs, the following experiment was done. Beginning April 10, ninety eggs each from the M and the B coops were gathered and stored in a dry room in sawdust under identical conditions. After 2, 4, and 6 months respectively, 30 eggs from each group were tested for their usability. It was evident in this testing that there was a very notable difference between the two groups, as shown in the table. The number of eggs spoiled in each lot of 30 eggs from each group, respectively, were:

	M	B
After two months	4 eggs = 14%	1 egg = 4%
After four months	14 eggs = 47%	6 eggs = 20%
After six months	18 eggs = 60%	8 eggs = 27%

Effects on Plant Growth

All farmers and gardeners know the importance of animal manure as a fertilizer. They know, too, that the dung of various animals—cows, sheep, chickens, and so on—affects plants differently. These differences affect plant size or rate of growth, as well as taste, quality of fruits and vegetables, and other aspects.

There is not just a difference between the manures of various animals, however; the manures of animals of the same sort show different effects according to the origin and quality of their food. Manure is a mixture of the decomposed products of elimination and the digestive juices, the latter being rich in hormones and ferments. Remaining in this are, to an observable extent, some of the qualities in the animal's food products. Hence, the effect of manure on plant growth also depends (aside from the health of the animal as such) on the fodder of the animal. *Whatever feeds well manures well* is a traditional adage.

There is even an observable difference in the effect of manure from animals that were given the same kind of feed, but that was grown in different ways. A difference can also be seen when the manure used is given special treatment—for example, when it has been biodynamically prepared. The following test shows a difference resulting from different feeding regimes.

Growth Test with Bush Beans
Average values for 12 plants in each case, in terms of 1 plant

	M	B	Control
Length of sprout in cm.	49	49	42
Weight of plants exclusive of beans gr.	58	77	44
Weight of beans: pods more than 10 cm.	79	83	29
Weight of beans: pods less than 10 cm.	4	9	6
Number of beans: pods more than 10 cm.	14	15	7
Number of beans: pods less than 10 cm.	7	11	11
Length of pods in cm.	14.7	14.5	12.2

Taste of Beans
Tested by 8 people; judgments were unanimous

	Raw	Cooked
M	very watery	good
B	firmer, more character	best
Control	dry, insipid	poor

On February 2, 1933, the manure from the chickens in the M and B groups was gathered and stored separately, and then mixed with the same sort of soil and set up in separate compost piles. Both compost piles lay side by side in the same conditions of light and warmth, and on March 14 had the same biodynamic preparations inserted in them at the same time.

Garden beds were fertilized on July 24 with these chicken-manure composts. One bed was fertilized with M compost, the other with B compost, a third was left unfertilized as the control C. The plant seed, sown in the three neighboring beds, was of exactly the same sort, and was allowed to grow under ordinary conditions.

Radishes and the previously mentioned bush beans were used as experimental plants.

The preceding and following tables outline these experiments, in every case showing the same result and differences, not only in quantity but also in quality, as was shown in the Havor test.

A like qualitative difference between the two chicken-manure compost heaps was also evident when the same experiment was made by "the method of sensitive crystallizations."*

Growth Test with Radishes
Average values for 16 plants in each case, in terms of 1 plant

	M	B	Control
Leaf length in cm.	12.6	9.5	9.6
Leaf weight in gr.	4.1	5.0	4.0
Number of leaves	4.5	5.6	5.6
Length of root in cm.	2.3	2.6	2.2
Weight of root in gr.	7	8.4	6.3

Taste: M: medium

B: strong, not sharp

Control: mild, too dry

An important relationship was discovered here. An original quality of the fodder—its nourishment value—continues on and influences the capacity for work of an animal; but in turn the quality of the manure itself is influenced and produces again new growth and new enhanced qualities as the seed, feed, and fertilizer cycle continues. Thus, the biodynamic cycle leads to a constant improvement in all the elements concerned until the highest degree of the performance and health is reached in every organism involved.

It is interesting, moreover, to note that animals react exactly to the variations in the quality of food. Grazing cattle select certain

---

*   Pfeiffer, *Formative Forces in Crystallization.*

plants and herbs and avoid others.* Farmers who know their animals also know what those animals like to eat. They see how cows
avoids plants of the crowfoot family (*Ranunculus*), which are poisonous in while green. They are afraid of the white hellebore and
meadow saffron, because they don't lose their poisonous quality,
even after being cut in the hay. Meadow foxtail (*Alopecurus pratensis*), orchard grass (*Dactylis glomerata L.*), and meadow fescue
(*Festuca pratensis*) are all well liked by cattle, while the broom grass
(*Andropogon glomeratus*) and the meadow sorrel (*Rumex acetosa*)
are eaten unwillingly or not at all.

With sure instinct, pasture animals choose the plants most beneficial to them. When farmers are sowing a pasture, they try—based
on their own or others' observations—to bring together those plants
that are gladly eaten by the animals. Owing to economic conditions
and to the state of science today (consider, for example the widespread, year-round stabling of animals), it is not always possible for
individual farmers to cultivate such intimate observations, which
are often replaced today by scientific experimentation.

This issue might be clarified by the following experiments, which
use white mice—the generally preferred as test animals because of
their sensitivity. In mice, the senses of taste and smell are extraordinarily developed. They are extremely nervous and sensitive animals
whose organs react strongly to changes in nourishment.

The experiments were done in such a way that the animals had
two similar dishes of wheat before them. This wheat was of the same
variety and grown for five years on the same soil for experimental
purposes, but the wheat in one dish was treated with mineral fertilizer
(calcium nitrate and ammonia plus Leuna® potassium nitrate), while

---

* Complete reports of these experiments are in Pfeiffer, "Vergleichender
Fütterungsversuch mit mineralisch gedüngtem und biologisch gedüngtem
Getreide," *Demeter*, vol. 6, no. 5; Pfeiffer and Sabarth, "Vergleichende
Fütterungsversuche mit Hühnern I," *Demeter*, vol. 7, no. 2; Pfeiffer and
Sabarth, "Vergleichender Fütterungsversuche mit Hühnern II," *Demeter*,
vol. 9, no. 1; Grone-Gültzow, "Einfluss der Düngung auf die Futterquali-
tät."*Demeter*, vol. 6, no. 9.

the wheat in the other was treated with biodynamic manure. The wheat in both cases was used in quantities that allowed the same percentage of nutritive substances. Basis: 12 tons of manure to the acre.

Experiment A was made with Bavarian brown wheat, and experiments B through E were done with Karsten no. 5. By weighing the uneaten feed daily, it was determined for each day how much of the mineral-fertilized grain the animals had eaten, and how much of the biodynamically fertilized. To obtain indisputable results, it was necessary to pay attention to several things that might seem to be of secondary importance. The two dishes of food were set up in exactly the same amount of light. Positions of the dishes were often interchanged to avoid the argument that the animals might always choose either the left or the right dish simply out of habit. To avoid spilling or losing any grain, neither the cage was too small nor the number of animals too great. By careful inspection and care of the cages, it was possible to keep all such mistakes to a minimum. The grain used was harvested under exactly the same conditions, so that it was all at exactly the same state of dryness. As supplemental food, we used boiled milk mixed with water in a 1:3 ratio for all the animals. If these points are observed, exact results can be expected.

The experiments were made either with animals that had, for 6 or 7 generations, been previously fed entirely on biodynamically fertilized wheat, or with other animals that had been fed only mineral-fertilized wheat. Table A shows such a selection test with the biodynamically fed female animals:

Table A	Biodynamic wheat	Minerally fertilized grain
6 Generations of biodynamically pre-fed mice eat in 24 days	108 gr.	10 gr.
6 Generations of minerally pre-fed mice eat in 24 days	98 gr.	5 gr

In this experiment, it quickly becomes clear that biodynamic feed was eaten almost exclusively. The animals tried the mineral wheat only at the beginning and then left it untouched. It could be argued that the animals were used to a certain type of food through inherited habit if an even more one-sided result had not shown itself in a family of mice that had been completely nourished "minerally" for six generations. Those animals completely quit their habit of eating mineralized food, and instead immediately chose only biodynamically fertilized grain when given the opportunity. The other kind of grain was only occasionally eaten.

The fact that an overbalance of salts in mineral fertilizing can have a harmful effect on food has been shown by various studies. The observations of Dr. von Grünigen has already been mentioned (pages 47–48). They point to the danger of a too much potassium content, which comes from a tendency of the plant to consume too much potassium. The data of Professor Rost go further in presenting the danger of a surplus of potassium.* He demonstrated in feeding experiments that, through the feeding of potassium, thrombosis and gangrenes could be produced experimentally. From this noteworthy treatise the following is cited:

> But now, in connection with the animals fed potassium nitrate, I made an extraordinarily interesting observation, for they showed a pronounced inclination in successive generations toward thrombosis.

Obviously, these phenomena appeared more pronounced in the second generation than in the mother animals. Rost states that thrombosis has also recently increased in human beings, up to 4 times its earlier prevalence, and he reaches the following conclusions:

---

* Prof. F. Rost, "Über Schwanz-und Fußgangrän bei Ratten," *Münchener Medizinische Wochenschrift*, no. 22, 1929.

The potassium content of plants can be considerably increased by potassium fertilizing.... In cooking, most of it goes into the water. Using the figures in König's *Chemie der Nahrungs- und Genussmittel** [Chemistry of food and beverages], one can reckon that with spinach, for example, about two-thirds of the mineral substances pass into the cooking water and is poured out with it.** But lately there is a strong tendency not to pour off this cooking water as was formerly done, but to use it, since these mineral substances are considered especially important for human nourishment. Of course, we absorb considerably more potassium salts with this modern way of cooking than previously, to which the fact must be added that, thanks to the plentiful use of artificial fertilizing, the potassium content of plants is higher than in earlier decades.

I may, perhaps, interject here, that this modern way of cooking in no way agrees with everyone's digestion. I know many otherwise completely healthy people who are affected by nausea and severe diarrhea as the result of eating vegetables prepared and the cooking water retained, and it is further known that the pollakisuria [abnormally frequent urination] so often observed during the [First] World War was mostly a polyuria caused by potassium salts.***

It is now natural for us to underline these sentences and conclude that the increased tendency to thrombosis, as we have observed it in recent years, has a direct relationship of cause and effect to the increased potassium content in food. I am personally of the opinion that we should be very careful about

---

*   *Chemie der Nahrungs- und Genussmittel*, IV Aufl., Bd. 2, S. 1458.

**   There was contained in:

    1. 100 gr. spinach, raw and unwashed 0.695 gr. potassium.
    2. 100 gr. spinach, raw but washed 0.0602 gr. potassium.
    3. 100 gr. spinach, cooked, cooking water poured off once, 0.192 gr. potassium.
    4. The water poured off from no. 3 contained 0.379 gr. potassium.
    5. 100 gr. spinach, cooking water not poured off, 0.601 gr. potassium.

    Thus, cooked spinach from which one does not pour off the cooking water contains about three times as much potassium as spinach prepared by the old method.

*** Unpublished research.

drawing such a conclusion in such a definite form.... We can indeed always say that, according to the animal experiments referred to and other data given on potassium feeding, such a conclusion is completely tenable and justified when considering the whole thrombosis issue from this viewpoint. Until now we have been unable to give any well-founded explanation for the increase of thrombosis in recent years. Hence, we can certainly consider this an encouraging result of the animal experimentations cited and be confident that they indicate a direction to follow in our researches into this highly important question.

Furthermore, these experiments are of interest because it was possible—through doses of salts that appeared harmless to the animal itself—to bring out conditions of sickness in the second generation.

The experiments of Tallarico,* too, prove the influence of fertilizing on the quality and the health-giving properties of foods. Note the following:

I was able to make certain observations in a series of experiments with grain that dealt with fertilizing and developing the mother plant, as well as the capacity for the yield of seeds thus produced. There was a noticeably different behavior in seeds coming from mineral-fertilized mother plants compared to seeds from mother plants fertilized with stable manure. While the first group in general gave a modest yield, the second, which was cultivated under identical conditions of soil, climate, and agricultural procedures, gave a more luxuriant growth and a higher yield.

Turkeys are very well suited for this sort of experimentation, since they eat almost anything and mature quickly. Above all, they are useful in this way because their time of puberty brings the so-called red crisis, at which time the characteristic

---

* G. Tallarico: The biological value of the products of soil fertilized with animal or with chemical fertilizer. "Proceedings of the R. Accademia Nazionale dei Lincei, Mathematical, Natural–Scientific Division," vol. 8, series 6, 1, Rome, Feb. 1931.

red growths appear on the bird's head and neck. During this critical period of development, the animal enters a state of profound weakness, so that it easily succumbs to intestinal or lung infections, which means trouble in even the best flocks. This natural sickness, which occasionally spares the strongest, comes somewhat early, lasts for a while, and ends in death or recovery, according to the attacked animal's powers of resistance. To estimate those powers of resistance, an experiment was made in which the number of ill birds in each test group was observed, as well as when the crisis began, its length in terms of days, and the manner of its termination. In the experimental flock there were also instances of birds that, as a result of a real and genuine recovery from the disease, remained small throughout life; in other cases, they remained so greatly weakened that they had to be removed from the flock because they were unable to obtain their share of food. These birds were given the special category of "stunted." The following is the report of the experiment:

The food was raised on 3 parcels of land. The *first parcel* was mineral-fertilized each year for 2 years with 180 pounds of ammonium sulfur-nitrate per acre, spread during the time of field cultivation, as well as 90 pounds of potassium per acre and 350 pounds of super-phosphate per acre that were spread after seeding.

The second parcel was fertilized each year for 2 years with decomposed stable manure in a proportion of 1,000 pounds per acre.

The third parcel was used in its natural condition without mineral or animal fertilizer.

On these 3 parcels, the most important foods for raising turkeys were grown in separate beds. In the second year, a portion of the produce grown on the same parcel the first year was used as seed. The foods thus grown were supplemented with egg yolk and ground meat and formed the 4 feed groups that were given to the 4 groups of experimental animals in the first 6 months of their lives in the form of a mash composed of equal parts by weight of the various ingredients:

Type A: meat residue, plus egg yolk, plus grits from stable-manured grain,* plus whole, stable-manured grain, plus stinging nettle and cut leaves of sweet clover, both the latter from parcels that had neither synthetic fertilizer nor stable manure.

Type B: meat residue, plus egg yolk, plus grits from mineral-fertilized grain, plus mineral-fertilized whole grain, plus stinging nettle and sweet clover leaves from unfertilized plants (fed with mineral-fertilized grains).

Type C: Meat residue, plus egg yolk, plus grits from unfertilized grain, plus unfertilized whole grain, plus stinging nettle and cut leaves of sweet clover—both from parcels that had been fertilized with stable manure (fed with stable-manure green feed).

Type D: meat residue, plus egg yolk, plus grits from unfertilized grain, plus unfertilized whole grain, plus stinging nettle and cut sweet clover leaves from parcels that had been fertilized with mineral fertilizer (fed with mineral-fertilized green feed).

To determine the organic capacity for resisting disease, the following phenomena were observed and considered: the number of sick animals, the beginning of the crisis for each animal, its length and its termination, whether in death, cure, or stunting. From these results, the mean percentage was then determined for each group and each test series, taking into consideration the unavoidable accidental losses in raising them—e.g., death through cold, injuries, or birds of prey.

From this it is clear that the turkeys fed in the first two months with grains or green feed from plants grown on stable-manured soil show, when they are attacked by the crisis, fewer cases of sickness, a shorter duration of it, and fewer cases ending fatally compared to the corresponding group of turkeys that, under the same conditions of life and environment, are fed with green feed or grains from plants that were fertilized with mineral fertilizer.

---

* Mineral-fertilized, stable-manured, and natural grains were given to the animals as crushed grains in the first month and as whole grain in the second month, the time when young turkeys begin to take whole grain.

In one table are gathered the average mean percentages obtained in each group for the three experiments:

Organic powers of disease resistance in young turkeys

Feeding with	% in crisis	Age in days at start	Dura-tion in days	% that died	% cured	% stunted
Stable-manured grains	89	43	7	21	79	0
Minerally fertilized grains	94	47	11	34	64	2
Stable-manured green feed	82	39	6	18	82	0
Minerally fertilized green feed	96	46	10	39	60	1

Furthermore, it can be gathered from this that the leafy growth of plants fertilized with stable manure has a more beneficial influence on the various phenomena considered here than do the reproductive organs of plants that are similarly fertilized with only stable manure. *This means that the seeds and, even more, the leaves of plants fertilized with stable manure have the peculiarity, when used as food for these animals, of increasing their capacity to resist disease more effectively than the corresponding seeds and leaves of mineral-fertilized plants.* The former thus have a higher biological value than the latter. This conclusion is also confirmed by the lack of cases of arrested development and stunting in the groups that were nourished with products fertilized with stable manure.

This presents an extraordinarily important aspect of the biodynamic method of agriculture. It is not only able to improve the soil's organic structure; its consequences also reach far into the human kingdom. If in only one instance the influence of various methods of agriculture on animal and human health as presented here is observed, then it is in the interest of every consumer to be concerned with the sources of the food one eats. We can and should demand of farmers that they furnish us with the maximum of health-giving qualities in our bread, vegetables, and fruit. The consequences for hygiene and health of such a stand on the part of consumers are incalculable.

Experience in this respect has always shown that wherever the biodynamic method of agriculture has been used, the attention of physicians is soon directed to the efforts of the products. Hence, we will cite several reports in this connection. The reader may see from them that agriculture is not a concern of only those who cultivate the soil, but that all human beings should be concerned about the way they nourish themselves.

J. Schulz, MD,* testifies that it was possible for him (with the help of biodynamically raised food in the form of bread) to cure a series of metabolic disturbances, and on the basis of this diet obtain a stronger effect with his medicinal therapy. He observed these beneficial results in children, as well as in adults. R. Reinhardt, MD,** and J. Kalkhof, MD,*** offer similar observations.

Another physician states:

> Biodynamic products are necessary for therapeutic control of diets and cannot be replaced by other products in the food market. As a physician, I acknowledge gladly that, especially for weak and challenged children, such food is necessary....
>
> Gradually, we have switched to using the biodynamic products, which seem to be of good quality and have a definite influence on the functions of the stomach and intestines. I have recommended these products to patients with marked stomach troubles and sluggish intestinal activity, and they have been fortunate enough to get over those ailments without medication....

---

* "Diäterfahrungen mit Demeterprodukten der biologisch-dynamischen Wirtschaftsweise," in *Fortschritte der Medizin* 7.1.35. Berlin.

** Reinhardt, "Einiges über Ernährung unter Berücksichtigung des Zusammenhanges van Ackerboden, Pflanze, Tier und Mensch," *Hippokrates,* vol. 5, no. 10, Stuttgart.

*** Kalkhoff, "Beobachtungen und Krankenerfahrungen mit Demeter-Ernährung," *Ärztliche Rundschau,* no. 21, 1935.

My broad experience as a dietitian with numerous patients has convinced me that, especially with a raw-food diet, the biodynamically treated products are preferable in every way to those that have been manured in the usual way with chemical fertilizer or by the use of feces.*

Physicians' Reports on Their Use of the Biodynamic Products: Cases A, B, C, D

A. "The excellent results I've had in using biodynamic products—in connection with the treatment of an ever-increasing number of my patients—have persuaded me of the urgent need today of having an unlimited supply of these products that have medical hygienic value for the general food supply."

B. "I am glad to state my appreciation of the value of the 'Demeter products.' I view them as a model to be followed at a time when the inner value of many of our foods is declining. It seems to be necessary, in the interest of public health, to have these products first on the markets, then later to introduce them into general use for everyone."

C. "I have no opinion concerning the cultural background of these products, but I endeavor to promote their use because of their beneficial effect on digestion. I try to advise my patients suffering from certain forms of digestive disturbances to use them. By using them I have been able to provide cures without drugs in many cases."

D. "I do not hesitate to state that, in my opinion, biodynamic products have the greatest value for public health—especially in diseases requiring dieting. My experience as a dietitian with numerous patients persuades me that especially raw vegetables and fruit grown by the biodynamic method are more healthful than the products of other methods of agriculture.

---

*   Unpublished; original belonged to the author.

In every case, it was observed that, with a diet changed to biodynamic products, there was first an immediate increase of appetite, which during the first weeks resulted in an increase in food consumption. After 2 to 3 weeks, however, a state of equilibrium occurs, with the final result that the person has enough to eat with only two-thirds the volume previously believed necessary. This fact is also a proof of the increased nutritive value of these foods. Extensive observations in this field have been made with hundreds of test cases over a period of years.

The well-known physiologist Privy Councilor Emil Abderhalden takes the following position with regard to the problem:

In connection with various human and animal illnesses it has frequently been desirable to trace them back to the means of fertilizing the food plants. Nothing can be said yet with any certainty, but we must bear in mind that important substances come from soil bacteria, and we must consider whether it is correct to disturb the fine interplay of all the soil organisms by bringing in nitrogen in the form of potassium nitrate and using lime and phosphoric acid, because the development of the various sorts of organisms is thus disturbed and hindered, and difficulties will someday arise because of this.

Elsewhere he says:

If we take care of the soil exclusively with chemical fertilizer, it is indeed conceivable that disturbances in the growth of plants will occur. It is especially conceivable that the development of unknown substances (vitamins) might be endangered. That organism—the soil—will certainly find itself in the same condition as an animal that receives nourishment substances only in their chemically purest form. Of course, it becomes ill. This organism—the soil, with its concourse of cells and their manifold reciprocal effects—will definitely become sick.*

---

*   Cf. the previously cited article of Dr. Reinhardt.

Before leaving this subject, let's cite an English research report.* R. McCarrison, B. Virwa Nath, and M. Suryanarayana are coauthors of a noteworthy study on the influence of chemical and organic fertilizing. They found important qualitative differences in the case of seeds of the millets *Eleusine coracana* (finger millet) and *Panicum miliaceum* (common millet) and of wheat. The differences were traced as far as feeding tests. Under the influence of the warm climate these grains give greater yields with organic fertilizer than with chemical fertilizer or without any fertilizing. In contrast to the yield without fertilizing, chemical fertilizing brought a yield increase of 32.8 % and organic fertilizing an increase of 100.7% in the case of *Panicum miliaceum*. The same strain of seed was used again and again for the same fertilizer test. Thus, an increase was attained in the various qualitative effects. Feeding tests with *Eleusine coracana*, using pigeons, gave the following results:

Average bodily weight loss during the test	
Group with basic ration	37.7%
Group with basic ration plus plants grown on stable manure basis	22.4%
Group with basic ration plus chemically fertilized plants	37.4%
Group with basic ration plus unfertilized plants	40.9%

Even with manifold changes in the conditions of the experiment, a better result was evident in the case of the seed raised with organic fertilizers than with seed raised with chemical fertilizers. In the case of the wheat, the seed raised with chemical fertilizers reacted less favorably than the seed raised without fertilizer.

----

* Virwa Nath, Suryanarayana, McCarrison, "Memoirs of the Department of Agriculture in India," 9, no. 4, 1927; cf. F. Dreidax, "Das Mark der Landwirtschaft," *Demeter*, no. 12, 1934.

Feeding Tests with Pigeons and Barley (loss in body weight)

Stable-manured barley showed better results than
"unfertilized" by 18.5%

Stable-manured barley showed better results than
"mineral-fertilized" by 15.0%

Experiments with rats, with a basic ration of meat residue, refined starch flour, olive oil and salt, cod liver oil (or Marmite or Vegex) for a vitamin supplement, and besides the foregoing, either organically fertilized wheat or mineral-fertilized wheat, both sorts grown on adjoining parcels, gave the following results:

Percentages of gain in bodily weight

Basic ration + stable-manured wheat 114%

Basic ration + chemically fertilized wheat + vitamin supplement . 104%

Basic ration + chemically fertilized wheat alone 89%

In other words, stable-manured wheat is even better, despite the sharp competition of a vitamin supplement.

We are surprised that these important experiments are thus far rarely quoted in scientific literature. It is assumed that such experiments run counter to many fondly held dogmas in the areas of agriculture and nutrition, but that should not prevent repeated objective examinations of this question.

When we also consider in this connection the researches of Professor Boas of Munich—who proved that organically fertilized pasture grasses have a higher albumin content than mineral-fertilized grasses,* which nonetheless contain more peptone (the first product of the disintegration of albumin)—we are justified in thinking that, when these facts are more widely known, people manifest a greater interest in the problem of quality in agricultural products than is now the case.

---

* Prof. F. Boas, "Untersuchungen für eine dynamische Grünland-Biologie," *Praktische Blatter für Pflanzenbau*, 9, 173, 1932.

CHAPTER 13

Practical Results of the Biodynamic Method

Theory and principle might be sound and correct, yet we must always ask: *Do they work well in practice?* In this scientific age there is yet another question: *Does the biodynamic method prove itself in scientific experimentation?* Scientific experiments have one peculiarity—it's goal is to be as precise as possible. Hence, all experimental errors must be eliminated insofar as possible. This requires narrowing experiments to encompass as little as possible, while retaining the functioning power of the principles to be tested.

If we are dealing with a complicated process such as plant growth in relation to soil and climate conditions, it is difficult or nearly impossible to keep an accurate account of all the factors. Thus, we are generally satisfied in such experiments to set up and test a series of detailed minutiae, and then assemble the whole from the parts. The question is whether we are then still connected with reality— i.e., are we still dealing with practical matters? We have done this sort of thing in testing the biodynamic method.

Elsewhere, in reports of scientific experiments, we have presented the results of one or another such experiments, by means of which a small segment of a whole problem could be shown. Nevertheless, it is generally recognized today that, speaking metaphorically, the most exact knowledge of what occurs within an internal combustion engine is in no way enough to enable one to build a truly good gasoline engine.

In an attempt to get closer to agricultural practice, we have directed the scientific method of experimentation toward something—i.e., the soil—that seems to be within the realm of actual practice. Thus, we performed "comparative experiments" in this field. Adjacent parcels of land were repeatedly cultivated in the same way, planted with the same crops, but fertilized differently to demonstrate the variations in fertilizer qualities. Such experiments were then carried out for as many as 4 years. In the past, the author himself has started and arranged such experiments, but was forced to recognize that they provide only an inconclusive picture of the biological alteration of the soil. Nonetheless, such a biological transformation is the very foundation of the practical value of this new method of agriculture. Such experiments cannot be conclusive unless the soil has not been rendered incapable of biological activity as the result of previous abnormal treatment.

Biologically enlivening the soil, changing its structure, and even improving certain qualities in it all require time. The more any of its factors has strayed from its natural basis, the longer is the time needed to return to normal healthy conditions. But this period doesn't belong to the experiment itself, but only to the preparation.

The "experiment" doesn't really begin in agricultural research until we get back, in the course of the crop rotation, to the same fertilizing and the same crop on the same land on which it was planted before. Then the soil has had time to develop, and the seed from the first crop can be used for the second. Then we know the peculiarities and special requirements of that particular soil, so that we can cultivate it in a way that the life stimulated in it can really come to expression.

The author once carried out a comparative experiment that was discontinued after three years as having been negative in its results. In that experiment, certain essential points had been ignored: a) the influence of a crop rotation; b) the fact that the experiment should not have actually begun until the third year after the field had been converted; and c) the special fact that for decades the land had been used for military field exercises. It had not been turned into farmland until 10 years earlier. Because this soil had been an exercise field for decades, it needed many years before it could be returned to a "normal and natural" state. In other words, we would have to consider the time needed for the ground to be properly prepared.

In another case, the experimental plot had previously been used intensively for several years for mineral-fertilizer experiments. Here again, there would have been a lag of several years before a normal state could be established.

We have frequently observed that the biodynamic seeds used in experiments have been shown to be more resistant to plant diseases—and that there is an increase in the effect under observation when there are numerous consecutive repetitions of an experiment in the same place.

Aside from all this, it is clear that the biodynamic method of agriculture does not merely result from fertilizing the soil. When properly practiced, biodynamics gives due weight to every factor needed for healthy plant growth. From this, we see why fragmentary tests, using comparative, parallel strips of land, are one-sided and thus do not lead to useful results—i.e., results are indecisive.

The ideal experimental basis is really a practical farm on which the results are checked and controlled over a period of years. Such a foundation, while it does include all the sources of experimental error found when dealing with nature instead of artificial groups of factors is, at the same time, the only correct one, because it considers the full and normal effects of nature, in contrast to an imbalanced, compressed nature.

This might seem complicated at first, but it is the only way everything can be done thoroughly and compared precisely. Here we can produce conditions that really have significance for practical application. If an exact means of crop rotation is determined for certain fields where the dung of the biodynamically fed animals is brought to the biodynamic parcel of land or the dung from the minerally fed animals is brought to the mineral-fertilized parcel, and if the products of these parcels are then given to the animals belonging respectively to these parcels, so that a closed cycle of substances is gradually obtained, then from practical experience we can judge, from all its effects, whether a certain method of agriculture has value or not.

Curiously, however, it is the scientists themselves who have a great hesitancy in attacking this problem, which is indeed a decisive one for the practical farmer or gardener. In other words: *What is a method's value when used on a farm over a period of years?* To answer this question, we present extracts of reports from a number of precisely observed and controlled farms, with figures on yields before and after their conversion to the biodynamic method.

Farm A: 450 acres of heavy and medium clay soil—grain and sugar beets the main crops—was intensively mineral fertilized before conversion. The conversion took place during 1922 to 1924 without the more recently accumulated experience for carrying it out.

The following table shows the usual crop rotation before the conversion. Manure was always given to the beets and potatoes.

	4 year		3 year		3 year
A	1st yr. beets	B	1st yr. beets	C	1st yr. potatoes
	2nd yr. wheat		2nd yr. wheat		2nd yr. oats
	3rd yr. beets		3rd. yr. rye		3rd. yr. rye
	4th yr. wheat		1st yr. beets		1st yr. potatoes
			2nd yr. wheat		2nd yr. wheat
			3rd yr. rye		3rd yr. rye

It took about 5 years to make the changes on this farm, one of the first to be worked completely biodynamically.

Crop succession after the conversion

4 Year

A. 1st yr. sugar beets plus biodynamic stable manure
2nd yr. field peas or beans plus oats
3rd yr. wheat
4th yr. rye or winter barley, with legumes (yellow clover)
immediately following

5 Year

B. 1st yr. potatoes plus biodynamic stable manure
2nd yr. wheat
3rd yr. oats plus field beans
4th yr. clover
5th yr. rye, followed by legumes

Average grain yield in pounds per acre

1914 – 1,746	1927 – 1,764
1915 – 1,817	1928 – 2,117
1916 – 1,834	1929 – 2,205
1917 – 1,755	1930 – 2,117
1918 – 1,975	1931 – 2,090
1919 – 2,006	1932 – 2.416
(conversion begun)	1933 – 2,160
1923 – 1,587	1934 – 1,605
1924 – 1,323	(long, heavy drought)
1925 – 1,675	1935 – 2,090
1926 – 1,675	

Average pea harvest before the conversion, in pounds per acre
(peas before conversion were an uncertain crop):   529–882
Average pea harvest after the conversion: 1,764–2,028
Average bean harvest after the conversion: 2.998–3,351

One argument frequently voiced by opponents of the biodynamic method is that things might indeed go well for several years, but the soil will nevertheless be exhausted after a longer period.

In 1932, 10 years after the conversion was begun, soil samples were sent to a scientific institution. Its report stated, "The soil is, at least at this time, sufficiently provided with potassium as well as with phosphoric acid."

1934: *"Abundantly* provided with potassium,
    abundantly provided with phosphoric acid."
1935: "The soil is *abundantly* provided with potassium,
    well provided with phosphoric acid."

Sugar beet yield in pounds per acre

1923 – 18,230	1930 – 26,638
1924 – 22,154	1931 – 26,898
1925 – 27,710	1932 – 27,714
1926 – 22,308	1933 – 28,248
1927 – 23,693	1934 – 31,740
1928 – 24,914	1935 – 25,029 (drought)
1929 – 25,025	

There was constantly a slight observable increase in the sugar content compared to the normal mean yield. This result has been repeated on various farms.

1927 sugar content of the biodynamic beets	17.2%	Mean figures for the sugar beet factory	15.27%
1934 sugar content of the biodynamic beets	18.2%	Mean figures for the sugar beet factory	18.14%

Potato yield in pounds per acre

Before		After	
1917	10,423	1931	20,370
1918	8,907	1932	17,637
1919	7,337	1933	17,637

Consumers praised especially the keeping quality and flavor of the biodynamic potatoes.

Milk production in pounds per year per animal

1914/15	5,924	1928/29	7,412
1915/16	6,433	1929/30	7,114
1918/19	4,877	1930/31	7,421
1919/20	6,541	1931/32	7,811
1924/25	6,581	1932/33	7,114
1925/26	6,938	1933/34	6,219*
1926/27	8,245	1934/35	6,429
1927/28	8,208	1935/36	6,479

There was an average of 30 to 35 cows kept on the place; they were fed mainly with fodder grown on the farm itself. Until 1928, 7 pounds of peanut meal and sprouted malt per animal per day were brought in from outside; until 1934, 3½ pound per animal per day; 1937, 1½ pound per animal per day. The farm had 1 head of cattle for every 5 acres of used arable land.**

Special emphasis is placed on feeding the cattle with home-grown fodder. We have seen that a mixed farm (with cattle and tilled fields) has greater lasting powers of production and of resistance to harmful effects than any other. Fortunately, forced feeding of cows to attain a record production of milk has gone out of fashion again. Natural cattle husbandry with meadows and hay and clover will always form the foundation of healthy farms; it furnishes the manure needed for field culture, while field culture in turn provides a portion of its produce to the animal husbandry. The "closed circle of feeding" (home-grown feed fertilized with home-produced

---

* Abortion (bang's disease) brought in from outside and slowly eradicated from the herd. 1936 shows further progress after the conclusion of this table.

** "Die Entwicklung des Klostergutes Marienstein bei Nörten" (Hannover), *Demeter*, no. 7, 1936.

manure) is the basis for a healthy stable. The fewer cattle replacements and the more home-bred, the better.

We encounter today, however, breeding diseases, contagious abortion, mastitis, and so on. These epidemic diseases are so familiar that we need only refer here to their existence. Nevertheless, we must listen to testimony from the field of practice.*

Farm A. "At the time we took over the farm, contagious abortion [brucellosis] had been present there over a period of years. Through the planned conversion in feeding and the animal husbandry in general, and with the help of the Weleda remedies, we were able to eradicate the contagious abortion from the stable. In the past year, our only loss was one calf, from diarrhea. All the cows calved normally."

Farm B. "The herd had been suffering miscarriages for 15 years, and were nearly annihilated. At this point, the biodynamic method of agriculture was introduced—the feeding plan was basically rearranged; pasture facilities were enlarged; the fodder was almost all raised on the place; and the feed-growing areas were put under intensive biodynamic cultivation. The new feeding basis and the treatment with Weleda remedies made possible an early, complete cure of the herd."

Farm C. "Regularly in the fall, at the time of winter stabling, and in the spring both calf paralysis and pneumonia were present. Losses ran as high as 30%. For the past two years, the farm has been worked biodynamically. Now the two diseases mentioned have practically disappeared."

Farm D. "Before conversion there was a great deal of sickness in the stable of this farm that, ironically, had belonged for three years to a member of an Association for Combating Breeding Diseases. It was also difficult to get the cows to calf. Until the time of conversion, there had been scarcely any improvement in the situation; after the conversion and treatment with

---

* Dr. N. Rehmer, "Erholzte Milch- und Fettleistung durch individuelle Betriebsgestaltung," *Demeter*, no. 1, 1937.

Weleda remedies an improvement soon began. Since then, the cows have been easily bred."

Farm E. "...until, in 1931, contagious abortion was prevalent. This drove us, in the course of the year, to weed out all 'purchased' cattle. There are now just a few of their offspring left. Since beginning with the biodynamic method of agriculture here, we decided to reform our herd entirely with local 'red and white' cattle. Our own breeding, since the conversion, has been very satisfactory. We give credit to the change in method that, since then, there have been no more cases of abortion. Since then, the fertilization and the carrying time of the cows has returned to normal. Breeding diseases are absent."

Farm F. "As is the case in other biodynamic farms, the first phenomenon was a luxuriant legume growth. Exceptionally good clover and harvests of 3,600 pounds of field beans and 2,500 pounds of vetch per acre were gathered. Primarily, there was improvement in the cow stable. Although the purchase of forcing feed was completely given up, milk production with only home-grown feed was raised by about 2,000 pounds per cow, and the butterfat content by about 12%. Our total figures on feeding showed that, with the same amount of 'starch units,' after conversion we obtained about a fifth more milk and fat than previously. These figures proved to us that there is a greater value in biodynamically raised fodder plants than in those grown with the customary agricultural practices. We feed the very last leaf of the fodder turnips without any sign of diarrhea. Our neighbors do not feed any turnip leaves, or very little, because they have found there is danger of diarrhea from their use. The general health of the herd has also improved most satisfactorily. During the past year, we were able to raise 20 calves from 21 cows."

We now present a series of examples of milk production before and after the conversion. The figures, with two exceptions, are taken from the reports of professional milk-production testing organizations.

Milk production in pounds per cow

	Year	Pounds of milk	Percent of butterfat	Remarks
1.	1929–30	9,116		Strong epidemic abortion-delayed calving
	1930–31	8,197		
	1931–32	6,420		⅔ of the fields converted to the biodynamic method
	1932–33	5,849		
	1933–34	7,460		
	1934/35	10,498		
2.	1914–24	4,409–6,614		Before conversion, production never exceeded 6,614 pounds
	1924–25	6,581.06		After conversion to biodynamic agriculture
	1925–26	6,938.56		
	1926–27	8,246.28		
	1927–28	8,207.72		
	1928–29	7,411.09		
	1929–30	7,113.36		
	1930–31	7,419.80		
	1931–32	7,811.44		
	1932–33*	7,113.36		
	1933–34	6,220.16		Drought
	1934–35	6,429.60		
	1935–36	6,478.94	3.40	
3.	1929–30	9,341	3.60	
	1930–31	9,643	3.27	
	1931–32	9,156	3.43	
	1932–33	9,480	3.28	Conversion to biodynamic method
	1933–34	1,0701	3.36	

* In recent years, the herd has been increased with its own heifers. The lowering of the average age of the herd and the greater proportion of young cattle, with a drought, helped to lower the annual average. It is, of course, well known that cows, in their first lactation periods, do not yet have their full production.

	Year	Pounds of milk	Percent of butterfat	Remarks
	1934–35	9,899	3.16	Pasture dried up
	1935–36	10,827	3·23	
4.	1931–32	6,354		In process of conversion
	1932–33	6,607		
	1933–34	7,418		Conversion completed
	1934–35	6,867	3.59	Drought
5.	1932	6,173	3.28	Conversion
	1933	6,393	3.24	
	1934	7,496	3.17	
	1935	7,055	3.11	Drought; on other farms in the local production-testing association, average production frequently dropped about 1,000 pounds
6.	1930/31	6,265	3·92	Conversion; there were still important changes in the number of animals during 1931–1933
	1931–32	5,747	3.95	
	1932–33	5,456	4.11	
	1933–34	6,196	3.71	
	1934–35	7,789	3.84	
	1935–36	8,527	3.73	

	Year	Pounds of milk	Pounds of butterfat	Remarks
7.	1932	8,541	274.25	
	1933	8,269	270.28	Conversion
	1934	9,028	293.87	
8.	1928–29	4,696	179.84	
	1929–30	5,567	210.84	Conversion
	1930–31	6,213	228.61	
	1931–32	5,811	218.43	
	1932–33	6,620	251.03	
	1933	6,662	248.61	
	1934	5,683	204.12	Drought

	Year	Pounds of milk	Pounds of butterfat	Remarks
	1935*	5485	203.19	Drought
			Percent of butterfat	
9.	1928–29	8,016	3.77	Conversion begun
	1929–30	8,091	3.73	
	1930–31	8,882	3.67	
	1931–32	9,156	3.81	
	1933	8,792	3.81	
	1934	9,321	3.62	
10.	1929–30	6,257	3.72	
	1930–31	5,710	4.06	
	1931–32	6,530	4.20	
	1932–33	7,639	4.21	
	1933–34	6,629	4.37	
	1934–35	7,513	4.08	Drought
	1935–36	7,756	4.18	
11.	1928–1935	7,344	3.09	Before conversion
	1934–35	8,327	3.11	After conversion
	1935–36	9,658	3.21	Before change used average of 980 lbs. linseed-oil cakes per cow; after, only homegrown grain and hay

Note: On this farm the effect of the new fodder quality was recorded in terms of figures. Before the conversion, 100 parts of starch units were transformed into 169 parts of milk. After the conversion, 100 parts of starch units were transformed into 215 parts of milk.

All the farm records quoted here are in Central Europe. There the biodynamic method of agriculture has been in use for the longest time. Hence, we are better able to make a comprehensive survey. The farms referred to are located in the most varied conditions of

---

* The number of cattle was increased by about 20%, hence the total milk production mounted.

soil and climate, and include places on plains as well as others in hilly country. The averages of milk production given are those of the complete herds. The annual average of 16 tested biodynamic farms in Germany was 7,366 pounds, while at the same time the annual average for all Germany (all farms) was 5,113 pounds, and that of all the tested farms (with 10% of the total head of cattle of the country) was 7,275 pounds. The data gathered in Germany apply correspondingly to the conditions of other countries in which there are biodynamic farms in operation.

We have purposely chosen the milk-production figures, because they give a decisive indication of the state of health of the cattle. This state of health, together with the manure produced by the cattle, represents the backbone of a biological agriculture. If the herd is in good condition, the fertilizing can be brought into good condition and the fertility of the soil maintained.

Worthy of special mention is an estate farm situated east of Berlin.* It is located in the least advantageous conditions of climate and soil. On a purely sandy soil in hilly country, surrounded by poor pine woods, this farm had rain as follows (given in millimeters):

1932	418	1935	436
1933	348	1936	330 (Jan.–Oct.)
1934	374		

Dew and mist and subsurface water are equally scarce. This farm has 150 acres of cultivated land and 37 acres of meadows. The soil is 70 percent pure sand. The biodynamic method was introduced along with the cultivation of mixed crops.

---

*   Dr. N. Rehmer, "Die Rindviehhaltung im Mittelpunkt eines Betriebsorganismus," *Demeter,* no. 10, 1936.

1	a	Potatoes plus 11 tons prepared stable manure per acre
	b	Oats plus lupin or oats plus vetch plus yellow clover (or summer rye plus summer barley plus sweet clover)
	c	Rye plus serradella
2	a	Lupine
	b	Rye plus 7 tons prepared manure per acre; followed by lupine
	c	Oats
3	a	Serradella or lupine grown for seed
	b	Rye plus 7 tons prepared manure; followed by lupine
	c	Oats or barley
4	a	Turnips plus 14 tons prepared stable manure
	b	Summer barley plus clover
	c	Clover
	d	Rye or spelt (wild wheat, triticale, spelt) followed by lupine
5	a	Oats

The mixed crops, with a generous use of legumes, markedly increased the soil fertility, so that after a few years even oats and clover gave sure yields. Before the conversion, the succession was potatoes, rye, summer grains, and much mineral fertilizer. By using mixed crops, a better ground cover was effected, with beneficial results such as an increase of moisture reserve and the shade it gave the soil. The increase of humus substances in the soil also helped its retention of moisture. In addition, this regime provided the fodder necessary for the manure supply.

The capacity of the soil for growing clover was increased, which showed itself, after a few years, by the fact that alfalfa did well there. Today they keep 2½ acres planted in alfalfa and 6¼ acres in clover. Potato acreage was decreased. Because of the cultivation of mixed feeds, it has now been possible to give green fodder in the stable from May to July. Clover is cut twice, alfalfa up to 3 times; it is pastured from August onward. Feeding is divided as follows: May to July, green-cut rye mixed with clover and alfalfa; July to October, pasture on serradella, clover, alfalfa, and on grass meadows, and supplementary feeding on sunflower seeds. November, sunflower seeds, mixed hay, turnip leaves, straw. December to April, turnips, straw, hay. Regarding the soil conditions, it must be noted that the souring of the soil was held in check not by spreading limestone on the ground, but by careful application of the sum of all the biodynamic measures.

In the course of several years, the herd, which had been up to 70% sterile, was cured. After the conversion to biodynamic procedures:

1932	from 23 cows	22 calves
1933	from 18 cows	17 calves
1934	from 17 cows	16 calves
1935	from 23 cows	22 calves
1936	from 23 cows	23 calves

On this farm, too—as we have found almost universally in our experience—it was again shown that the purchase of registered cattle from noted breeding establishments does not pay. The development of strong types, suited to local conditions, progresses slowly but surely.

We cannot give all the details concerning this farm, which is of such special interest because it was run under such unfavorable conditions in the beginning. We shall, therefore, only add certain other figures. One of the cows on the place had, in 15 years, 12 calves of which 5 have already grown up to be valuable cows. Her milk production is today, in her fifteenth year, still more than 9,920 pounds. Of her bull calves, three were used for breeding purposes (see tables on page 187).

A markedly strong bone development was noted in the bull calves. If we keep in mind that the development of the farm took place on a sandy soil, this speaks more emphatically than any other fact for the practical usability of the new method.

As an example, in the field of intensive gardening, we present the yields—which have remained the same for six years—of an enterprise

Average milk production of the farm per cow per year in pounds

Before the conversion	No. of tested cows	Milk in pounds
1926–27	10	5,401
1927–28	11	7,401
1928–29	10	6,063
After the conversion	No. of tested cows	Milk in pounds
1932	20	7,055
1933 ⎫	16	7,297
1934 ⎬ drought years	13 lower age of herd	7,385
1935 ⎭	14	7,716
1936	20	7,275

We also cite the individual development of certain young animals:

Year	Pounds of milk per year				
	B	C	D	E	F
1932	8,157	6,360	6,380	3,931	—
1933	9,094	6,516	6,788	6,111	5,414
1934	11,005	7,571	8,488	6,614	6,991
1935	11,354	7,668	9,524	8,602	7,745

in Holland. A greenhouse of about 2,000 square yards, planted every year with tomatoes: 3,000 plants produce from 8 to 9 tons (cf. ch. 9). In this connection it is to be noted that this yield has come evenly for six successive years from the same piece of soil, and without plant diseases. A greenhouse with cucumbers—40 yards long, 3.75 yards wide—produces each year about 4,400 green cucumbers. One of similar size produced about 2,700 white cucumbers annually. A grape hothouse, 40 yards long and 8 yards wide, yielded, beginning with the third year, about a ton of grapes (Frankenthaler variety).

Let's consider a few more experiences on larger farms. Dr. von Heynitz writes that, before converting to the biodynamic method of agriculture, he paid out an average of 50 Reichsmarks per hectare

(2½ acres) for chemical fertilizer, and about 70 to 80 Reichsmarks per hectare of land for additional concentrate feeding.* He was unable to get yields corresponding to this outlay, so he started experimentally converting a part of his place and then converted the entire farm of 286 hectares (715 acres). He states:

> I was particularly well able to evaluate the crops at this time, since in the first three years of the conversion a part of the fields was still chemically fertilized, while to an increasing extent the other section of the farm received biodynamic treatment. I was able to prove, in comparing yields, that the yields of the biodynamically fertilized crops were not quantitatively behind those produced with chemical fertilizer.

An important phenomenon, observed everywhere, is that biodynamic grain shows less tendency to "lodge" under moist conditions.**

After the conversion the yields were:

Yields in pounds per acre after the conversion		
Rye	2,363	
Winter wheat	2,985	
Summer wheat	3,144	
Winter barley	3,092	
Oats	2,637	
Potatoes	21,032	(18–22% starch, depending on the variety)
Sugar beets	20,106	(18.8% sugar)
Fodder turnips	77,690	

These data concern a farm in hilly country with a heavy soil, in Saxony. The use of concentrated feeding is interesting; after the conversion, it cost 17 Reichsmarks per hectare as against 70 previously. This was done while maintaining the milk yield at a point of 6,614 to 7,716 pounds per animal per year with 3.25% butterfat.

---

* Dr. B. von Heynitz, "Meine Erfahrungen mit der biol. dynam. Wirtschaftsweise und dem Absatz ihrer Erzeugnisse," *Demeter,* no. 2, 1934.

** To "lodge" means to stay down after being beaten down.

From another report, concerning an intensive grain-growing farm on a light clay and silicious soil, we take the following data.* We are dealing here with a definitely poor farm. Two earlier tenants went bankrupt, a third paid no rent, a fourth was barely able to get along. The conversion to the biodynamic method proceeded slowly in the course of 5 years:

Yields

Year	Total acres	Biodynamically treated acres	Yield in pounds per acre
			Wheat
1924–25	165	—	1,834
1925–26	107	—	1,640
1926–27	210	—	1,852
1927–28	107	—	1,799
1928–29	115	—	2,840
1929–30	130	—	2,183
1930–31	215	27	1,481
1931–32	157	110	1,949
1932–33	140	140	2,319
			Rye
1924–25	157	—	2,266
1925–26	87	—	1,675
1926–27	97	—	1,808
1927–28	200	—	2,169
1928–29	107	—	2,390
1929–30	112	6.3	2,319
1930–31	72	10	1,684
1931–32	95	32.5	2,290
1932–33	152.5	152.5	2,205
			Winter barley
1924–25	—	—	—

*   Dr. A. Vogelsang, "Der Betrieb Rittergut Böhla," *Demeter*, no. 2, 1934.

1925–26	20.3	–	1,182
1926–27	13.3	–	2,884
1927–28	17.5	–	2,690
1928–29	40	–	2,346
1929–30	55		2,857
1930–31	70	–	2,610
1931–32	67.5	6.2	2,593
1932–33	67.5	67.5	2,699
			Oats
1924–25	95	–	1,808
1925–26	220	–	2,801
1926–27	120	–	2,222
1927–28	150	–	1,870
1928–29	132.5	–	2,328
1929–30	100		2,028
1930–31	82.5	–	2,399
1931–32	57.5	–	1,870
1932–33	62.5	57.5	1,808
			Potatoes
1924–25	95	–	10,847
1925–26	145	–	12,875
1926–27	122.5	–	13,757
1927–28	127.5	–	8,871
1928–29	122.5	6.2	12,346
1929–30	142.5	32.5	13,510
1930–31	140	140	13,424
1931–32	**125**	125	15,450
1932–33	122.5	122.5	15,609
			Sugar beets
1928–29	6.2	–	23,300
1929–30	10	–	32,804
1930–31	15	–	23,810
1931–32	15	15	20,635
1932–33	15	15	21,870

Considering weather differences during individual years, the fluctuation and occasional decline in yields are not of major significance, since soil tests that were carried on simultaneously with keeping crop records showed no decline of nutritive materials in soil.

A small farm of 23 hectares (57.5 acres) on medium and light soil, in a region poor in rain, reports:

| | Average yield in pounds per acre | |
	Before conversion	After conversion
Wheat	1,360	1,640
Rye	960	1,160
Oats	1,560	1,720
Barley	1,520	2,000
Corn (maize)	2,360	3,000
Fodder beans	1,920	1,840

It is a general experience that small farms can be converted more quickly and simply than the very large ones. Smaller farms are generally fertilized intensively—that is, they have more cattle per acre, as the case just cited. Here the careful, biodynamic handling of the manure has especially beneficial effects that explains the relatively high yields.

These experiences, gathered under continental European conditions, are directly applicable for England and America, too. It is the author's firm conviction, based on personal studies during various visits to England and America, that the standard of agricultural work in these countries could, by means of careful work, be greatly increased. This would, of course, require the cooperation of many farmers. The proper procedure in this connection would be first to set up, in various locations in both countries, model and demonstration farms. These could then be visited by all farmers. On these farms, the adaptability of the methods to the most varied conditions of climate and soil could be demonstrated. It is a familiar truism that farmers learn more through practical demonstrations than from any amount of literature, because they are more persuaded by what they see than by what they read.

The goal of this book, therefore, is to constitute a brief instruction in how to make "being able to see the thing in practice" possible. This is not a textbook or instruction for farming; rather, it presents long-neglected views that will, when put into practice, restore conventional farming to a healthy condition, which will in turn help to foster a permanently sound agriculture and a healthier human society.

CHAPTER 14

Human Responsibility

European agriculture is on the threshold of a transition from traditional methods to conscious farm management. Yet, Europe-
ans are still standing on "solid ground," which allows nearly biologically balanced agriculture, and in this lies the chance of a hopeful
future. It is true that biological possibilities in Europe that the land
will have already been overstepped, but in a way that still permits
turning back and a renewal.

Europe, as a middle region between East and West, represents
a kind of center of gravity in the previously described tendencies
of the Far East and the Far West. Here, life conditions exist from
which the reestablishment of a natural basis for food supply can be
developed. Here, methods might be devised in a healthy way, and
biological laws recognized that would be beneficial to the people of
Europe—indeed, to the whole West, as well as to the diseased condition of the earth organism in the East. Here, outer nature as it were
extends a helping hand.

However—and here we come to a point that might seem surprising to many—the solution does not depend on nature but on human
beings. The solution of the present agricultural crisis is a human
spiritual problem. It consists of people extending their knowledge of
nature's being, life's laws, and the *creation of a method of thought
based on the principle of an organic whole.*

If the foundations of traditional culture are lost—a process that
is becoming more evident in all areas of life as a fundamental trend

of the twentieth century—then nothing will help, not turning this way or that, not probing and not overly intellectual calculations, discoveries, and applications. All study and all that we might do remain a mere patchwork so long as the one great task—the creation of a new culture—is ignored and not recognized.

If this new work we do does not become a mere appendage of the 1,002nd case to the 1,001st, but instead a means of gathering and presenting a body of knowledge imbued with life and the capacity to develop, then we can find a way out of our difficulties. If we provide a particular food, grow a specific plant variety of grain—indeed, if we do something that, in all its details, fills our day as agriculturists—it will all remain patchwork if these efforts are not based on a fundamentally different attitude on our part toward the problems of life and growth—an attitude that allows us to perceive life and growth as an organic whole over the entire earth.

Some will argue that this is all very well but has nothing at all to do with agriculture as such. The answer is that the one-sidedness of prevailing views of life, nature, and the universe have been one of the primary reasons for the collapse of our culture, as the events of our times themselves [in 1930s Germany] fully demonstrate.

Others will say: *Very well, we acknowledge the broad, constructive advance implied in your first, incomplete presentation of this perspective; let's go ahead and rearrange everything on that basis.* The optimum response is that the best lecture on soil treatment will not help at all when attempting to apply what is said if the farmhands working in the fields do not know how to plow. Metaphorically, today we have too many "lectures" and not enough "farmhands." With this in mind, the following facts may be briefly restated.

Seeds need time to develop their qualities. They are planted repeatedly according to the rhythms of crop rotation. In agriculture, we know that when the same crop is planted in the same plot of ground for 3 to 7 or more years (i.e., long periods), only then does the value of crop rotation become clear. We learn to observe and work in terms of long rhythms of development. A forester's rhythms stretch out even further in time. The goal—even when clearly known and established—can be attained only in the course of the rhythm of development. This is the first basic truth: *Curing the ills of agriculture cannot be attained today or tomorrow, nor can it be done in a few years.* Since this issue is concerned with processes of growth, the rhythms and times of these aspects must be noted and strictly followed. Those who work with a plan to return their individual farms back to health know and take this into account.

The first step on such a farm is to gather and carefully nurture all organic fertilizer material that might be on hand. This alone is the basis for producing humus in the soil.

The second step is proper cultivation of the soil. If humus already exists, it is necessary only to maintain this. The important thing is correct, soil-conserving crop rotation. Although the first step can be completed in one year, the second requires a period of 4 to 8 years.

The third step is to improve the cattle, for they in turn furnish the "raw material" for the first step. This means that repetition of the first step—gathering organic fertilizer—has more meaning later on, because it takes place at a higher level of quality, since it is produced by the farm's own improved cattle.

The fourth step is to biologically reshape the farm's environment, its biological area as a whole. Presumably, in the meantime we have gathered sufficient wisdom for this through observation and experience.

Keeping step with the process of development on individual farms and applying the principle that offers a living understanding of a whole region of country (after we have tried it out in miniature) does not mean finding a solution for any single difficulty, but working so that the program as a whole is furthered and made healthy. Perhaps the most significant words in Sir Merrick Burell's resolution are *"Only a carefully thought out, long-term agricultural policy!"* In other words, *don't start by producing tractors if there's no one to operate them,* as one hears in [Soviet] Russia.

Goethe and Rudolf Steiner are milestones on this path of development. Goethe enunciated, for the first time, consciously and in harmony with exact thinking, the laws of the organic, superior whole and the method of cognition belonging to it. Rudolf Steiner carried the Goethean idea further until, finally, it is in a form that practical agriculturists, the farmers, can grasp.

Elsewhere, this volume describes practical applications of the biodynamic method of agriculture, a method that includes, to begin with, coming to understand the laws that govern our perception of the living, functioning life processes; in other words, we have been "Goethean" in developing our ideas.

Farmers who wish to convert their farm according to biodynamic views must first work on themselves and learn to think in different ways. This is the greatest difficulty involved in introducing marked innovations; people would love to have an instant recipe for getting their work done without having to use their own inner activity. This is impossible, of course, in practical areas of life such as farming, gardening, and forestry. As human beings, we are the strongest natural force that guides and directs the beginning, middle, and end of the natural growth process; our capacity is the final, decisive factor.

We've known of farmer who began deepening their interest in what goes on in the fermentation of manure and compost and suddenly encounter most significant discoveries dealing with the formative forces in life. A good example of this can be seen in the way good bread is made; ripe manure makes crumbly soil; grain ripens properly on it. After mowing, it goes through a further ripening process in the ear; if it is threshed immediately after mowing, this ripening process cannot take place. Furthermore, the best quality is retained by storing the grain in the ear rather than as threshed seeds, and beetles and mice do less harm. We find that late-threshed grain ripened in the ear makes better bread. Threshed grain "sweats"—it is still alive—and "ripens" further, and it can be ground when it reaches a state of rest. It is then in the best condition for baking, although the flour continues to "work" and needs another 3 to 5 weeks to reach the peak of its baking value. Does anyone still recall these things? They were once common knowledge! Who considers these things now? How often are chemists and technicians called in to help fix defects arising from forgetfulness and ignorance!

Biodynamic farmers themselves are trained to notice the finest processes. But this presupposes that a slow, inner change is taking place within them. As they learn to recognize and comprehend the value and potential for developing the life process in a deeper sense through a "rotation" of the spiritual processes of perception, a real farmer gradually develops from a "soil-tilling mechanic" into a real farmer. An ethical feeling of responsibility develops toward that organism, "the living soil." Reverence for life as a whole is fostered—an inner relationship to one's calling as a "tiller of the earth" unfolds. Today, biodynamic farmers are inwardly upheld by conscious knowledge (formerly, it was instinctive tradition), but beyond this way of working is neither consciousness nor tradition. Herein lies the establishment of a new "peasantry." Such conscious knowledge is the only guide that can lead to healing the worldwide sickness of agriculture and society.

"Industrial" farmers and "rent chasers" become restless at this point. In any case, they have no idea of how to improve agricultural conditions. But, the notion that *this should take place by way of the inner spiritual attitude of individuals and the relationship toward their calling*—well, that is asking way too much! In the meantime, however, farms are being cultivated that have already shown, based on 10 and more years of experience, that the new point of view and approach has already affected farmers, who have managed to stamp out the "diseases" of their stable, and during dry periods their fields form a green island in the brown landscape. Above all, they are learning to love their soil again.

An inner and an outer transformation in harmony with the laws of life were the first results of the biodynamic method of agriculture. Numerous farms stand as witnesses to this—farmers who have done this work correctly. But any revolution in a farmer's habits of working and thinking must be decided by each one, individually.

If you were to question an experienced biodynamic farmer today concerning the solution of the world agricultural problems, the answer would have to be *no short-time program, but a plan that looks ahead for two generations; gradually turning over the land from incorrect to correct practices; and even before beginning to work the ground, the training of the those involved is needed.*

Those who want to work the soil will have to look around to find those with open-minded attitudes and skill and capacity for the conscious direction and shaping of the biological cycle. Such people—trained in even the smallest practical details—could go to individual farmers to live with them and exemplify, on the spot, the true farming life in its hard, daily effort and toil. This would be the first step in a plan that embraces perhaps two generations—adult education in its truest sense. If this problem is resolved, then bringing of the soil back to health is really already accomplished. Those who understand this already hold the key.

Farmer—in your hands lies the future!

CHAPTER 15

Summary

The Goethean attitude of mind—by perceiving the fundamental laws of a higher unity, a higher organism in every part, and by grasping the whole not simply as the sum of its parts, but also by seeing a comprehensive spiritual idea and world order become effective, and, with the "perceptive power of thought," by surveying both in their cooperative activity—indeed offers a way of knowing that endows people today and in the future with an ability to direct the events of world evolution on an orderly, harmoniously balanced course. Such knowledge requires only study, development, and earnest zeal to bear fruit. It takes courageous decisiveness to recognize and to effect this new and future knowledge without regard for the customary thought habits of the present whereby we strive for position and respect or, without consideration, for one-sided successes with no general economic and social value.

Yet here, too, natural development acts as an aid to knowledge. Not only as admonisher, but also as an aid and friend, nature steps up to the side of those who allow her to speak impartially and free of bias to their awakened spiritual organs. Chance observation and systematic search show the way.

The writer of these lines, whose interest was awakened by his study of Rudolf Steiner's Spiritual Science, has discovered ample evidence—which can presently, by the way, be observed by anyone—of the problems involved in the fertility of the earth described in this book. This earth fertility preserves itself wherever there is a harmonious distribution and relationship of forest, field, meadow, lake, river, and moor, wherever the soil is treated intensively to the highest degree—not to its maximum capacity but corresponding to its degree of fertility—and wherever manual labor is employed intensively.

Small farms, especially in regions that have developed a genuine culture and an instinctive knowledge of the soil through the centuries, are superior to farms worked intensively by industrial methods for the sake of its economic value and profitableness.

Within the limits of a certain region, if a small farm unit is present with the right proportions of tillable fields, meadows, pastures, and wooded areas, along with a balanced humus and manure distribution, then such a farm can still be treated by the owner in its entirety, and it will not only be profitable through the preservation of its "soil capital" (its soil fertility, which is soil capital), but also profitable in its yield value, economically and biologically. For example, plowing is an art. Those who survey their soil—which their family has preserved for generations—and know all its characteristics can determine the correct depth to plow, judge the right time to do it, and estimate the moisture condition to ascertain the soil's proper crumbly state.

Native farmers and farmhands who have been attached for years to a locality are superior to transitory migrant workers. On small farms, permanent attachment to the soil is stronger than on large farms or even in colonial settlements, among which we must count America. The writer of these lines has on various trips to America made a striking observation, confirmed by experts: Loss of humus, erosion, and so on occur least on small homestead farms—that is, in areas where real European farmers settled and continued their farming traditions with tools with which they were familiar. This applies to the Pennsylvania Dutch region and to areas where French and German farmers settled. This applies to the regions that border forested, hilly ranges in any way. This applies especially to mixed-crop farms with a balance between tillable land and meadows. This is an indisputable fact.

During periods of drought, we have found that such farms remain healthy and green. It has been reported to me, and as we already know, that despite exact investigations, the official agricultural bureau in Washington had for a long time no desire to know anything about these facts. Only gradually has the significance of the diversified farm begun to be appreciated. It is with this that the concept of the family farm, the homestead farm, is to be associated, the farm that can be preserved in the family for generations, because the peasant farmers (of Europe) have learned to preserve the fertility of the soil.

How this can occur has been described in detail by the author. The striking thing about this has been that such regions and farms demonstrate an economically propitious condition, in spite of economic crises, overproduction, and natural catastrophes. This fact is of great national economic significance. It affords a stable economic status with a modest but certain livelihood. The American situation described (we could equally well use an example from a European country) shows further that this sort of small, diversified farm—among the Pennsylvania Dutch, for example—held out through the economic crises and (as an expert described it to me) even fulfilled its tax and other obligations, and required no state or federal subsidy or help, while during the same critical period a subsidy of one million dollars was given to an immense monocrop sugarcane plantation in south Florida. It was one of the largest subsidies ever given to a single agricultural enterprise, my informant assured me.

To repeat:

First: Local soil fertility is achieved by means of the small diversified farm; by means of intensive humus–manure–compost economy (model: the biodynamic method); by preserving a regulated and balanced crop rotation; by increased use of human labor on the land; and by avoiding speculation on agricultural products.

Second: Fertility of a region is achieved through a harmonious distribution of field, forest, meadow, lakes, and rivers; by water economy and protection of the watershed; by afforestation; by careful drainage of swamps and marshes, with careful consideration of the underground water level and preservation of its circulation possibilities; by preserving forests and doing everything possible to develop a diversified reforestation (of clear-cut lands); through avoidance of monoculture; by avoiding real estate speculation (especially valuable agricultural land) for purposes such as industrial or colonizing projects; by establishing protective plant and forested areas against wind and overly intensive sun irradiation.

Third: The fertility of an entire country is preserved by the consideration of factors 1 and 2. In addition to these, however, are other problems. As is the case on individual farms and surrounding areas, everything depends on the right intermingling and diversification of plant growth, as well as the distribution of field, forest, meadow, and lake, all of which are even more important for an entire country. Here, things that should be studied most and brought into harmonious interrelationships are the key positions of water, wind, and, general climatic conditions. Wooded hill and mountain ranges and steep slopes as water collectors and dispensers should be protected against erosion by afforestation. Moderately steep and fertile slopes, as in China, can be partially protected by terracing; this work is an absorbent of superfluous labor power.

On the plains, the drying sweep of winds should be prevented by wood copses and hedgerows at the most frequent intervals possible. In this way, the monotony of the region will be broken and the minds of individual people, by viewing a pleasing landscape, will be stimulated to reflect on and appreciate nature. Flat countrysides, or plains, are thus not only an object of exploitation and profit, but on the whole offer an aesthetic perspective, lifting the mind to a sense of moral values, since it stimulates thought and reflection.

Let's study for a moment the difference between the mentality of mountain farmers or valley farmers, and that of farmers who cultivate barren "steppe" land or a settled flat region. This difference is immediately evident from purely outer considerations, such as styles of architecture, habits of life, rural festivals, and customs. If for a moment we compare various countrysides and note from which regions of a country the most spiritually productive and important human beings originate—its philosophers, artists, technically trained scientists, and others—our astonishment will be great.

The equilibrium of parts within the countryside means not only fundamental principles of evolution in harmony with the course of nature, but also means, similarly, the distribution of people themselves, which leads to both natural and social consequences. We described depletion of labor forces in rural districts. These are drawn together in the industrial centers of large cities. This sort of agglomeration exerts a powerful influence on climate, which has not yet, unfortunately, been sufficiently investigated and considered. The example of Berlin is well known, whose rainfall in the western parts of the city results in a "rain shadow" in its eastern areas beyond the main city.

The problem of water supply of large residential districts is well known; large population centers can disturb the water supply of an entire region. On the plains, where the countryside is flat and water is drawn from artesian wells or subterranean streams, they can even cause a drop in the water table. Poisonous smoke from industrial districts (above all, the sulfurous acid content) shows its effect on neighboring forests. The exploitation of oil wells consumes a considerable amount of subterranean water, allowing the water to flow back into abandoned well holes, whether intentionally or over the course of time.

The most important thing is the lack of work on land that, because of insufficient tillage, can lead in many places to the gathering of scarcely any harvest. Proper care of forests in the United States is neglected because wages are too high. In other words, labor is proportionately too dear. We can assume that industrial wages are an inducement to live in cities. As a result, industrialization acts destructively by dislocating the equilibrium between wages and living conditions. It drives people into two extreme groups, seeming to favor one while discouraging the other and harming both. This disproportion in its repercussions on the basis of life itself is one of the main causes of the deterioration described. Instead of helping, industrialization has become a human adversary, because it has made itself independent of natural laws of nature and necessary moderation, and without restraint has grown dissolutely rank. When the consequence of this path affected the labor market by causing unemployment, we were confronted by a surprising problem—that industrialized city dwellers no longer had the capacity for working on the land. Both voluntary and compulsory resettlement have failed, just as the immigrant settlement of nonagricultural populations has also frequently failed.

Here at the outset, we have a large educational task—reorientation of city dwellers so they might be equal, bodily and spiritually, to the task of tilling the soil and feeling kinship with the earth again. Knowledge of the phenomena of growth and development in the living Goethean sense will help substantially. Awareness of being responsible for the earth's fertility—i.e., bearing responsibility for the culture of the future, supported by technical knowledge and love of work—is the moral basis for this task.

Industry and commerce are estranged from nature and must, on their part, contribute something to reestablishing equilibrium. Regardless of how utopian this might sound today, this contribution should be made voluntarily through insight. If not, then catastrophes of nature and the socioeconomic life will compel people to do so. In this case, however, because of violent pendulum swings to extremes, the establishment of equilibrium cannot occur until the distant future, accompanied by incalculable suffering.

We would then see that, through economic limitations, such things as surplus and excess labor power, capital, economic initiative, and energy could be used to benefit the earth. In this we would see a worthy future for humankind. In any case, this process would be healthier than one that comes about through the correction and destruction of surplus values by wars and revolutions or by the orders of governments.

The practical aspect of this will mean that more people work on the land again. They will work more intensively. Methods for the preservation of humus will make possible an intensive cultivation of the soil. Individuals will live, as it were, in the midst of districts where horticulture is intensified. They will become members of village communities as the basic form of the social structure. Larger centers would be needed only for the administration of governmental and business affairs, schools, and so forth. Hasn't it been shown already today in many places that decentralized industries with small factories become better able to withstand crises—small factories whose labor forces own nearby lands and gardens (e.g., Württemberg), or who produce single parts of larger articles by working at home, where the factory is more an assembling plant (e.g., in Japan)? Workers, once more connected to the soil, become capable of a new social attitude.

Much of the regions that are presently becoming "steppes" could thus be transformed into "gardens." The examples are present; it is really the egoism of individuals and fixed habits of thinking among the masses that hinder this cultural state from arising.

Human habits in general play an important role in such matters—for instance, the habit of whiling away time and quieting mental unrest by reading newspapers for hours at a time. Let's examine this single habit; the amount of paper needed to print a single edition of a Sunday newspaper in New York City, for example, consumes 75 acres of forest wood. The question arises: How can this expenditure of nature be justified, even from the broadest and most liberal perspective in regard to the earth and human culture?

Fourth: In addition to the means already described, the fertility of an entire portion of this or another continent will be preserved through the reciprocal agreements of its inhabitants. Also, these people must become aware of their position in administrating the controlling factors of the whole climatic character and biology of their continent. The organization of a single region has significance not only for the inhabitants, but also for the inhabitants of neighboring countries. Deforestation of France would mean a prairie climate for northern Switzerland and Germany and erosion in the Alps. Deforestation of Russia would mean the spread of the tragic Asiatic conditions of life in the direction of Europe. During the winter, dust storms would whirl from the east westward. Drying winds could sweep across whole continents unhindered.

Let's compare, for example, the size and extent of high and low air-pressure regions in America and Europe. In Europe, there are greater variations and more frequent changes than in America. Among individual countries, a division of labor might be arranged on the part of life common to all—fertility of the earth. Indeed, so-called international relationships would receive a fuller, richer content for the first time as a result. These relationships offer a common task for the cultural life and future of humanity. The Goethean thought of incorporation of the individual parts—in the case of the life of a people, through insight and resolution and awareness of its position as a responsible part of the whole into the laws of a higher unity, or into "the entire further evolution of our cultural period"— can achieve here its ultimate realization: An *IDEA* whose reality of life assures the future of the human race!

Cited Works

This list does not include all books quoted in the text

Abderhalden, Emil: *Bisher unbekannte Nahrungsstoffe und ihre Bedeutung für die Ernährung. Halle A. S. Zeitschrift für Schweinezucht*, Heft 17, 1922.

———. "Nahrungsstoffe mit besonderen Wirkungen unter besonderer Berücksichtigung bisher noch unbekannter Nahrungsstoffe für die Volksernährung." *Die Volksernährung, Veröffentlichungen aus dem Tätigkeitsbereiche des Reichsministeriums für Ernährung und Landwirtschaft.*

Anthony, R. D. *Soil Organic Matter as a Factor in the Fertility of Apple Orchards.* Pennsylvania State College School of Agriculture and Experiment Station; State College, Pennsylvania, no. 261, Jan. 1931.

Bartsch, Dr. Erhard. *Die Not der Landwirtschaft, ihre Ursachen und ihre Überwindung*, Emil Weises, Dresden.

Blank. *Handbuch der Bodenkunde,* Bd. I bis VIII, 1929.

Bouyoucos, George J. Rate and Extent of Solubility of Minerals and Rocks under Different Treatments and Conditions. Agricultural Experiment Station of the Michigan Agricultural College. *Technical Bulletin,* no. 50, July 1921.

Demeter, "Monatsschrift für biologisch-dynamische Wirtschaftsweise." Bad Saarow, 1928–1937.

Dreidax, Franz. *Der Regenwurm,* aus Koschützki, *Rationelle Landwirtschaft,* Berlin.

Fagan, F. N. Anthony, R. D., and Clarke, W. S. Jr.: *Twenty-five Years of Orchard Soil Fertility Experiments,* The Pennsylvania State College School of Agriculture and Experiment Station, State College, Pennsylvania, no. 294, Aug. 1933.

Fippin, Elmer O. "Nature, Effects, and Maintenance of Humus in the Soil," *The Cornell Reading-Courses,* vol. 3, no. 50, The Soil Series no. 3, Oct. 15, 1913.

———. "The Soil: Its Use and Abuse," *The Cornell Reading-Courses,* vol. 1, no. 2, Soil Series no. 1, Oct. 15, 1913.

———. "Tilth and Tillage of the Soil," *The Cornell Reading-Courses,* vol. 2, no. 42, The Soil Series no. 2, June 15, 1913.

Gustafson, A. F. "Organic Matter in the Soil." *Cornell Extension Bulletin,* no. 68, New York State College of Agriculture, Cornell University, Ithaca, NY, Oct. 1923.

Howard, Sir Albert. *The Manufacture of Humus by the Indore Process.* The Royal Society of Arts, London, vol. 84, no. 4331, Nov. 22, 1935.

Howard, Albert and Yeshwant, D. Wad: *The Waste Products of Agriculture: Their Utilization as Humus.* Humphrey Milford. Oxford University, London, 1931.

Jenkins, S. H., Ph.D.: "Organic Manures." *Imperial Bureau of Soil Science, Technical Communication,* no. 33, Harpenden, England, 1935.

Jenny, Hans.: *Soil Fertility Losses under Missouri Conditions.* University of Missouri, College of Agriculture, and Agricultural Experiment Station, no. 324, May 1933.

———. *A Study on the influence of Climate upon the Nitrogen and Organic Matter Content of the Soil.* University of Missouri, College of Agriculture, and Agricultural Experiment Station, *Research Bulletin 152,* Nov. 1930.

Kallet, A., and F. J. Schlink. *100 Million Guinea Pigs.* New York, 1933.

King, F. H. *Farmers of Forty Centuries of Permanent Agriculture in China, Korea and Japan.* London, 1926.

Kapff, S. "Von der künstlichen chemischen Düngung zur natürlich-biologischen Wirtschaftsweise." *Demeter,* 10. Jahrgang, no. 8.

Liek, L. *Der Einfluss der Düngung auf die Zusammensetzung der Nahrungsmittel. Leib und Leben.* München, Aug. 1935.

Krafft: *Lehrbuch der Landwirtschaft.*

News Sheet of the Bio-Dynamic Method of Agriculture, no. 3, London, 1936.

Niklewski, B. *Der Einfluss der Kompostdüngung und Behäufelung der Pflanzen auf Ernteproduktion (Streszczenie),* Warschau, 1929.

———. *Zur Biologie der Stallmistkonservierung. Centralblatt für Bakteriologie, Parasitenkunde und Infektionskrankheiten. Band 75.* Jena, 1928.

Pfeiffer, Ehrenfried E. "Een nieuwe methode om betere landbouwproducten te krijgen." *Natur en Techniek,* no. 10, Oct. 1934.

———. *Formative Forces in Crystallization.* Rudolf Steiner Publishing, 1936.

Cited Works

———. *New Methods in Agriculture and Their Effects on Food-stuffs*. The Biological-Dynamic Method of Rudolf Steiner. London: Rudolf Steiner Publishing, 1934.

———. *Sensitive Crystallization Processes: A Demonstration of Formative Forces in the Blood*. Spring Valley, NY: Anthroposophic Press, 1975.

———. *Über den Einfluss van Kolloidstoffen auf die Entwicklung einiger Kulturpflanzen. Jahrbücher für wissenschaftliche Botanik, 1933, Band 78, Heft 3*. Verlag von Gebrüder Bornträger. Leipzig.

———. *Wind, Luft und Staub als bodenbildende Faktoren. Kalender 1934/35*, Mathematische–Astronomische Sektion am Goetheanum, Dornach, Switzerland.

Piper, C. V. and Pieters, A. J. "Green Manuring." USDA, *Farmers' Bulletin*, no. 1250, Washington, April, 1925.

Rost. "Über Schwanz- und Fußgangrän bei Ratten." *Münchener Medizinische Wochenschrift*, 76. Jahrgang, no.. 22, May 1929.

Schlipf. *Lehrbuch der Landwirtschaft*.

Schwarz, M. K. *Ein Weg zum praktischen Siedeln*. Pflugscharverlag Klein, Vater und Sohn, Düsseldorf, 1933.

Secrett, F. A. "Discussion: Proceedings of the Society." *The Journal of the Royal Society of Arts*, vol. 84, no. 4334, Dec. 13, 1935.

Slipher, J. A. The Management of Manure in Barn and Field, Bulletin of the Agricultural College Extension Service, Ohio State University, Columbus, OH, May/June 1914.

Stapledon, R. G. *The Land Now and Tomorrow*. Faber, London, 1935.

Sussenguth, A. "Aus den Grenzgebieten der Medizin; Pflanzenernährung und Volksgesundheit." *Deutsche Medizinische Wochenschrift*, Verlag Georg Thieme. 59. Jahrgang, no. 51, Leipzig, Dec. 1933.

Weir, Walter W. *Soil Erosion in California: Its Prevention and Control*. University of California, Berkeley. Bulletin 538, Aug. 1932.

Wiancko, A. T., G. P. Walker, R. R. Mulvey. *Legumes in Soil Improvement*. Purdue University, Lafayette, IN, no. 324, July 1928.

Wolfer. *Grundsätze und Ziele neuzeitlicher Landwirtschaft*. Verlag Paul Parey, Berlin.

Books in English by Ehrenfried E. Pfeiffer

Biodynamic Farming and Gardening: Renewal and Preservation of Soil Fertility. Hudson, NY: Portal Books, 2021.

The Biodynamic Orchard Book. Edinburgh: Floris Books, 2013.

The Biodynamic Treatment of Fruit Trees, Berries and Shrubs. Wyoming, RI: Bio-dynamic Farming and Gardening Association, 1957.

Chromatography Applied to Quality Testing. Wyoming, RI: Bio-dynamic Farming and Gardening Association. 1984.

Pfeiffer's Introduction to Biodynamics. Edinburgh: Floris Books, 2011.

Sensitive Crystallization Processes: A Demonstration of Formative Forces in the Blood. Spring Valley, NY: Anthroposophic Press, 1975.

Subnature and Supernature in the Physiology of Plant and Man: The True Foundations of Nutrition. Spring Valley, NY: Mercury Press, 1981.

Using the Biodynamic Compost Preparations and Sprays in Garden, Orchard, and Farm. San Francisco: Bio-dynamic Farming and Gardening Association, 2002.

Weeds and What They Tell Us. Edinburgh: Floris Books, 2012.

Further Reading

Berg, Peter. *The Moon Gardener: A Biodynamic Guide to Getting the Best from Your Garden.* Forest Row, UK: Temple Lodge, 2012.

Cook, Wendy E. *The Biodynamic Food and Cookbook: Real Nutrition that Doesn't Cost the Earth.* Forest Row, UK: Clairview, 2006.

————. *Foodwise: Understanding What We Eat and How It Affects Us: The Story of Human Nutrition.* Forest Row, UK: Clairview, 2003.

Groh, Trauger. *Farms of Tomorrow Revisited: Community-Supported Farms—Farm-Supported Communities.* Hudson, NY: Anthroposophic Press, 1998.

Grohmann, Gerbert. *The Plant: Volume 1: A Guide to Understanding Its Nature.* Hudson, NY: Anthroposophic Press, 1989.

————. *The Plant: Volume 2: Flowering Plants.* Hudson, NY: Anthroposophic Press, 1989.

Hill, Stuart B. *The Biodynamic Farm: Agriculture in Service of the Earth and Humanity.* Great Barrington, MA: SteinerBooks, 2006.

Keyserlingk, Adalbert. *The Birth of a New Agriculture: Koberwitz 1924 and the Introduction of Biodynamics.* Forest Row, UK: Temple Lodge, 2009.

————. *Developing Biodynamic Agriculture: Reflections on Early Research.* Forest Row, UK: Temple Lodge, 2000.

Klett, Manfred. *Principles of Biodynamic Spray and Compost Preparations.* Edinburgh: Floris Books, 2005.

Klocek, Dennis. *Sacred Agriculture: The Alchemy of Biodynamics.* Great Barrington, MA: Lindisfarne Books, 2013.

Koepf, Herbert. *Koepf's Practical Biodynamics: Soil, Compost, Sprays, and Food Quality.* Edinburgh: Floris Books, 2012.

König, Karl. *Nutrition from Earth and Cosmos.* Edinburgh: Floris Books, 2015.

————. *Social Farming: Healing Humanity and the Earth.* Edinburgh: Floris Books, 2014.

Masson, Pierre. *A Biodynamic Manual: Practical Instructions for Farmers and Gardeners.* Edinburgh: Floris Books, 2014.

Morrow, Joel. *Vegetable Gardening for Organic and Biodynamic Growers: Home and Market Gardeners.* Great Barrington, MA: Lindisfarne Books, 2014.

Osthaus, Karl-Ernst. *The Biodynamic Farm: Developing a Holistic Organism.* Edinburgh: Floris Books, 2010.

Philbrick, John and Helen Philbrick. *Gardening for Health and Nutrition: An Introduction to the Method of Biodynamic Gardening*. Hudson, NY: Anthroposophic Press, 1995.

Philbrick, Helen. *Companion Plants and How to Use Them*. Edinburgh: Floris Books, 2017.

Scharff, Paul W. *Commentary on Rudolf Steiner's Agriculture Course: From the Paul W. Scharff Archive*. Great Barrington, MA: SteinerBooks, 2018.

Selg, Peter. *The Agriculture Course, Koberwitz, Whitsun 1924: Rudolf Steiner and the Beginnings of Biodynamics*. Forest Row, UK: Temple Lodge, 2010.

Steiner, Rudolf. *Agriculture: An Introductory Reader*. Forest Row, UK: Rudolf Steiner Press, 2010.

———. *Agriculture: Spiritual Foundations for the Renewal of Agriculture* (CW 327). Kimberton, PA: Bio-dynamic Farming and Gardening Association, 1993.

———. *Agriculture Course: The Birth of the Biodynamic Method* (CW 327). Forest Row, UK: Rudolf Steiner Press, 2010.

———. *What Is Biodynamics? A Way to Heal and Revitalize the Earth*. Great Barrington, MA: SteinerBooks, 2004.

Strong, Devon. *A Lakota Approach to Biodynamics: Taking Life Seriously*. Great Barrington, MA: Lindisfarne Books, 2016.

Thornton Smith, Richard. *Cosmos, Earth, and Nutrition: The Biodynamic Approach to Agriculture*. Forest Row, UK: Rudolf Steiner Press, 2009.

Thun, Maria. *Gardening for Life: The Biodynamic Way*. Stroud, UK: Hawthorn Press, 2000.

Thun, Matthias. *The Maria Thun Biodynamic Almanac*. Edinburgh: Floris Books, annual publication.

———. *The Maria Thun Biodynamic Almanac: North American Edition*. Edinburgh: Floris Books, annual publication.

Trédoulat, Thérèse. *The Moon Gardener's Almanac: A Lunar Calendar to Help You Get the Best from Your Garden*. Edinburgh: Floris Books, annual publication.

Wildfeuer, Sherry (ed.). *Stella Natura Biodynamic Planting Calendar: Planting Charts and Thought-provoking Essays*. Pottstown, PA: Growing Biodynamics, annual publication.

Wright, Hilary. *Biodynamic Gardening: For Health and Taste*. Edinburgh: Floris Books, 2009.

About the Author

Ehrenfried E. Pfeiffer (1899–1961) began working with Rudolf Steiner in 1920 to develop and install special stage lighting for eurythmy performances at the first Goetheanum building in Dornach, Switzerland. Following Steiner's death in 1925, Pfeiffer worked in a private research laboratory at the Goetheanum and became managing director of an 800-acre biodynamic experimental farm in Domburg, Netherlands. The international work of testing and developing Steiner's agriculture course of 1924 was coordinated by Pfeiffer at the Natural Science Section of the Goetheanum. His most influential book, *Biodynamic Farming and Gardening,* was published in 1938 in five languages. The following year, only months before the outbreak of World War II, Pfeiffer hosted Britain's first biodynamics conference, the Betteshanger Summer School and Conference, at the estate of Lord Northbourne in Kent, considered the "missing link" between biodynamic agriculture and organic farming after Northbourne published his manifesto of organic farming, *Look to the Land,* which coined the term *organic farming.*

Pfeiffer first visited the U.S. in 1933 to lecture to a group of anthroposophists at the Threefold Farm in Spring Valley, New York, on biodynamic farming. His work afterward was essential to the development of biodynamic agriculture in the U.S.

Pfeiffer developed an analytical method using copper chloride crystallization as a blood test for detecting cancer. As a result, he was invited to the U.S. in 1937 to work at the Hahnemann Medical College in Philadelphia. During that time, he consulted with those interested in biodynamic farming and helped to form the Biodynamic Farming and Gardening Association in 1938. Escaping the advance of German troops into France, in 1940 he immigrated to the U.S. from Switzerland with his wife Adelheid and their children, Christoph and Wiltraud. In Kimberton, Pennsylvania, Alaric Myrin offered Pfeiffer an opportunity to create a model biodynamic farm and training program. While at Kimberton, Pfeiffer led the initiative to establish the Biodynamic Farming and Gardening Association and its journal. Later, Pfeiffer bought a farm in Chester, New York, where a small colony arose that focused on farming, education, and the administration of the Biodynamic Association.

In 1939, Pfeiffer's copper chloride sensitive crystallization theory brought him an honorary Doctor of Medicine degree from Hahnemann Medical College and Hospital in Philadelphia. He studied chemistry and became a professor of nutrition in 1956. He also wrote on the dangers of pesticides and DDT, and Rachel Carson consulted with him when she was writing *Silent Spring*.

In 1961, at his home in Spring Valley, Ehrenfried Pfeiffer suffered a series of heart attacks, but was not given proper medical care and died. His wife subsequently assumed responsibility for running their farm in Chester.

CPSIA information can be obtained
at www.ICGtesting.com
Printed in the USA
BVHW031539101120
592841BV00019B/364

9 781938 685293